Meeting Policy Challenges for a Sustainable Bioeconomy

OECD
BETTER POLICIES FOR BETTER LIVES

Please cite this publication as:
OECD (2018), *Meeting Policy Challenges for a Sustainable Bioeconomy*, OECD Publishing, Paris.
http://dx.doi.org/10.1787/9789264292345-en

ISBN 978-92-64-29233-8 (print)
ISBN 978-92-64-29234-5 (PDF)

Photo credits: Cover © Courtesy of NOVAMONT.

Corrigenda to OECD publications may be found on line at: *www.oecd.org/about/publishing/corrigenda.htm*.

Foreword

This book is dedicated to the memory of Mr Peter Schintlmeister (Austrian Federal Ministry of Economics, Family and Youth). We were deeply saddened and affected by Peter's death in January 2017. At the time, Peter was chair of the OECD's Working Party on Bio-, Nano- and Converging Technologies (BNCT). He was a stalwart of OECD efforts in industrial biotechnology for several years, and, known throughout Europe and beyond for championing the bioeconomy and the OECD's work in this area.

This publication represents the work of the former OECD Working Party on Biotechnology (WPB) and its associated task force, the Task Force on Industrial Biotechnology (TFIB), as well as the BNCT. The book was supported by many engagements with the public and private sectors and by the following 15 workshops, either OECD-led or jointly managed with other organisations:

- OECD: *Sustainable biomass drives the next bioeconomy: A new industrial revolution?*, Paris, France, 10-11 June 2014
- European Commission/OECD: *Present and future policy for bio-based production*, Turin, Italy, 9 October 2014
- Ministry of Science and Higher Education, Government of Poland/OECD: *Opportunities and challenges presented by synthetic biology*, Warsaw, Poland, 25 November 2014
- HUGO/OECD: *Genomics for sustainable development in emerging economies: Health care, food, environment and industry*, Kuala Lumpur, Malaysia, 14 March 2015
- OECD: *The long-term potential of marine biotechnology*, Plentzia, Spain, 29-30 September 2015
- Cluster SPRING/OECD: *Biowaste biorefinery: Biowaste exploitation in multi-purpose biorefinery schemes*, Rimini, Italy, 6 November 2015
- German federal government/OECD: *Bioeconomy policy analysis: Reconciling food and industrial needs*, Berlin, Germany, 26 November 2015
- European Commission/OECD/FAO: *Bioeconomy indicators*, Brussels, Belgium, 14 December 2015
- International Risk Governance Council (IRGC)/OECD: *Adaptive risk governance in synthetic biology*, London, United Kingdom, 8 January 2016
- OECD (TAD and STI): *Innovations in food and agriculture system: Policies to foster productive and sustainable solutions*, Paris, France, 25-26 February 2016
- OECD: *Innovation for a sustainable bioeconomy*, Paris, France, 25 May 2016

- OECD: *Public policy for a sustainable bioeconomy*, Glasgow, United Kingdom, 18 October 2016
- European Commission/OECD: *Biomass supply and demand for a sustainable bioeconomy: Exploring assumptions behind estimates*, Brussels, Belgium, 23 February 2017
- Tekes/Ministry of Economic Affairs and Employment of Finland/OECD: *New innovation ecosystems and circular solutions to boost the bioeconomy*, Helsinki, Finland, 7 June 2017
- Cluster SPRING/OECD: *Successful industrial examples of circular bioeconomy in Italy*, Rimini, Italy, 9 November 2017.

The BNCT Secretariat drafted the report with expert oversight of the WPB, the TFIB and the BNCT.

Table of contents

Tables

Figures

Boxes

Acronyms and abbreviations

AD	Anaerobic digestion
ADBI	Asian Development Bank Institute
ART	Artificial reproductive technology
BBI	Bio-based Industries Initiative
BCI	Biorefinery Complexity Index
BCP	Biorefinery Complexity Profile
BOD	Biochemical oxygen demand
BRICS	Biological Robustness in Complex Settings
CAGR	Compound annual growth rate
CEN	European Committee for Standardisation
COD	Chemical oxygen demand
DHA	Docosahexaenoic acid
EF	Environmental footprint
EFSI	European Fund for Strategic Investments
EIB	European Investment Bank
EPA	Eicosapentaenoic acid
ERC	Engineering Research Center
ERT	Economically relevant trait
ESIF	European Structural and Investment Funds
EURICS	Energy Union Integrated Research, Innovation and Competitiveness Strategy
FCI	Feature Complexity Index
FQD	Fuels Quality Directive
GH	Glycoside hydrolase
GHG	Greenhouse gas
GIFF	Green Investment Financing Forum
GM	Genetic modification
GMP	Good manufacturing practice
GSI	Genetic Stock Identification
ICSASG	International Cooperation to Sequence the Atlantic Salmon Genome
IEA	International Energy Agency
ILUC	Indirect land-use change

IPN	Infectious Pancreatic Necrosis
IRO	Intermediate research organisation
IUU	Illegal, unreported and unregulated
LC-PUFA	Long-chain polyunsaturated fatty acids
LiDAR	Light detection and ranging
LUC	Land-use change
MAS	Marker-assisted selection
MECs	Microbial electrolysis cells
MFC	Microbial fuel cell
MOOC	Massive open online course
MSW	Municipal solid waste
MTBE	Methyl tertiary butyl ether
NGO	Non-governmental organisation
NIBRT	National Institute for Bioprocessing Research and Training
NREL	National Renewable Energy Laboratory
PEFCR	Product Environmental Footprint Category Rules
PET	Polyethylene terephthalate
PHA	Polyhydroxyalkanoate
PPP	Public-private partnership
PRODIAS	PROcessing Diluted Aqueous Systems
QTL	Quantitative trait loci
R&D	Research and development
R&I	Research and innovation
RED	Renewable Energy Directive
RIN	Renewable identification number
RFS	Renewable Fuel Standard
SIRA	Strategic Innovation and Research Agenda
SNP	Single nucleotide polymorphism
SPIRE	Sustainable Process Industry through Resource and Energy Efficiency
STEM	Science, technology, engineering and mathematics
TFP	Total factor productivity
TRL	Technology readiness level
USDA	United States Department of Agriculture
US DOE	United States Department of Energy
US EPA	United States Environmental Protection Agency
US FDA	United States Food and Drug Administration
VFA	Volatile fatty acid

Executive summary

The bioeconomy concept has emerged from niche interest to political mainstream with over 50 countries publishing bioeconomy policies and intentions. It has also grown from a biotechnology-centric vision to an economic activity that spreads across several key sectors and policy families: agriculture and forestry, fisheries, food, trade, waste management and industry. As a result, the bioeconomy policy environment is much more complex than before. One intention of this book is to reflect that changing environment. It sets out what a bioeconomy policy framework might look like based on the familiar innovation divisions of supply- and demand-side policies. It brings up to date the science and technology implications for policy makers.

The bioeconomy concept envisages a gradual replacement of fossil-based feedstocks with bio-based ones. While this replacement implies an inherently more sustainable production system, this is not necessarily the case as organic resources can be overexploited. Bioeconomy policy, first and foremost, must ensure that biomass is collected and used in a sustainable fashion. Ensuring "biomass sustainability" implies living within the boundaries of what the planet can provide. It has come to mean much more than just environmental sustainability, and as such a sustainable bioeconomy also has to create jobs and wealth, and distribute that wealth more evenly.

When the grand challenges of climate change, food security and energy security are added to this mix, the situation becomes even more complex and relates to other large policy areas such as rural regeneration, re-industrialisation, the circular economy and smart specialisation. Bioeconomy policies must align with these larger goals in an efficient manner, with minimal overlap and duplication of effort and by wisely directing public funds. Policies must work across levels, which presents challenges and opportunities: biorefineries are effectively regional facilities and regulation is nation-based, while R&D is international and biomass trade is already global.

The book focuses significant attention on the institutional level, especially at key facilities: biorefineries and bioproduction plants. These are often cast as small- to medium-scale production facilities, often to be built in rural locations so they are closer to the biomass feedstocks. This is a major departure from the highly centralised, integrated oil refinery and petrochemical plant production model that is dominant in the fossil era. This alone has major policy implications: fragmented value chains, high-risk investments in biorefineries, and untried or non-existent industrial ecosystems that demonstrate the need for new forms of public-private partnerships.

Key messages

Biotechnologies, including industrial biotechnology and engineering (or synthetic) biology, remain a big part of the bioeconomy concept, offering great potential in this future vision. These can be regarded as platform technologies that span several key sectors – agriculture and crops, food and beverage, pharmaceuticals, chemicals and materials, energy and even national security. However, large technical obstacles remain as the cost of bioproduction is generally still too high. All too often research success is not accompanied by commercialisation. There are large skills gaps, and countries will continue to struggle with making and educating the bioproduction workforce.

Additionally, national bioeconomy strategies tend to demonstrate intent and commitment, but be short on detail, due in large part to the large range of related policy families, including tax, innovation, industry, agriculture, waste and trade. Experience shows that policy must take account of both supply- and demand-side measures, yet the latter, while a potential source of innovation, has tended to be overlooked by governments. Demand-side measures include public procurement, regulation, standards, consumer policies and user-led innovation initiatives. They also include lead market initiatives to address market and system failures in areas with pressing social needs. All should be seen as necessary components of a sustainable bioeconomy policy framework as supply-side measures alone are unlikely to build this future vision.

In many engagements with the bio-based private sector, the most consistent message is that bioeconomy policies have to be stable and long-term so that the private sector has the confidence to invest. One suggestion has been to have a 15-25-year competitive advantage over the fossil industry. Expensive as that may seem, fossil subsidies are still astronomically high, and climate change is real. Risk mitigation for the private sector goes beyond policy certainty, although the latter is a very important factor. Financial instruments for building public-private partnerships have to be attractive and not overly bureaucratic.

A carbon price and carbon tax seem like the logical way to raise the large sums required to finance the public contributions of such projects. Pricing carbon emissions through a carbon tax should be a powerful incentive to invest in cleaner technologies and adopt greener industrial processes.

Objections to subsidising young technologies of any sort for climate change mitigation can be based on arguments around market distortion caused by subsidies. However, there is no such thing as a "level playing field" between the fossil industries and any of the green industries – including industrial biotechnology and engineering biology, which are foundational technologies of a bioeconomy. The fossil industries are over one century old and fossil fuels subsidies are still gargantuan: therefore the argument seems hollow. Removing fossil fuel subsidies and pricing the environmental damage of those industries would put a completely different complexion on their economics, and would make arguments against green bioindustries much less convincing.

Finally, all is dependent on sustainability of the feedstocks, the processes and the products of a bioeconomy if the mistakes of the past are not to be repeated in the future. Biomass sustainability as a policy subject is extremely complex and cannot be resolved without international – if not global – support.

Chapter 1.

The bioeconomy concept: Then and now

The bioeconomy concept is expanding rapidly. Around 50 countries, including the G7, have either a national strategy or policies consistent with a future bioeconomy. While many published strategies have laudable goals for solving large societal problems, they lack policy detail. Moreover, the bioeconomy concept means different things in different nations. As a result, gathering comparable metrics is becoming a real challenge. For these reasons, a policy framework for a bioeconomy would be useful for countries to identify their relative strengths and weaknesses, fill policy gaps and understand the bigger picture for the international bioeconomy. This chapter provides an overview for such a framework.

Overview

There is no universally agreed definition of "bioeconomy". Consistent with OECD (2009), this report understands bioeconomy as the set of economic activities in which biotechnology contributes centrally to primary production and industry. This is especially the case where advanced life sciences are applied to the conversion of biomass into materials, chemicals and fuels. Nevertheless, policy must reflect that the bioeconomy has moved beyond biotechnology. It is in fact embedded in the far-reaching transitions that are taking place in energy, transport and industrial production (OECD, 2017).

Momentum has been building for the bioeconomy for over a decade. The OECD set the wheels in motion within the membership with its landmark 2009 publication, *The Bioeconomy to 2030: Designing a policy agenda.* Events of 2015 propelled the bioeconomy concept to the forefront of politics: the Conference of the Parties (COP21), the UN Sustainable Development Agenda and its 17 goals, and the Global Bioeconomy Summit. These events responded to the so-called grand challenges of climate change, energy security, food and water security, and resource depletion. However, the bioeconomy is aligned naturally with more mainstream policy, such as knowledge-driven reindustrialisation, circular economy, smart specialisation, green growth and rural regeneration.

The world has realised that economic growth can be allied to environmental policy goals via a bioeconomy. At least 50 nations (Figure 1.1), including the G7, have put in place national bioeconomy strategies or have policies that are steering towards a bioeconomy (El-Chichakli et al., 2016). Since then, France, Italy, Norway, Spain and the United Kingdom (at least) have produced or are working on dedicated bioeconomy strategies.

Figure 1.1. **How the world is gravitating towards bioeconomy policy**

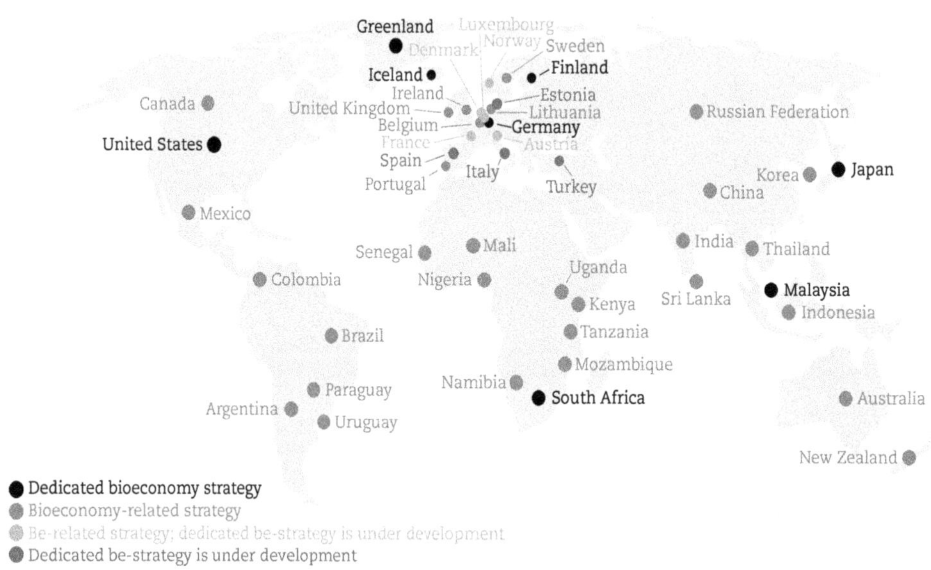

Note: be = bioeconomy.

Source: Adapted from Bioökonomierat (2018), "Internationale Bioökonomiestrategien", http://biooekonomierat.de/biooekonomie/international.

The transition to an energy and materials production regime based on renewable resources is expected to be fraught with many setbacks and obstacles, technically and politically. Earlier transitions from wood to coal and then from coal to oil were not complicated by the grand challenges faced today. Bennett and Pearson (2009) argued the transition from coal-based to petrochemical feedstocks in the United Kingdom occurred between 1921 and 1967. However, they pointed out the transformation was not inevitable. It was hastened by mass production of cars in the United States in the 1920s. More or less by the end of the 1940s, the United States had a large supply of olefins for transformation to petrochemicals. Diffusion east took time, but by the late 1960s the UK organic chemical production industry was totally transformed to petrochemistry.

Bioeconomy policy makers can take at least one lesson from this: the transformation to a bioeconomy will take time. The energy transition is at least two decades old already and is proving expensive: the cost of *Energiewende*, recently described as "Germany's energy gamble" (Schiermeier, 2013), is expected to top EUR 1 trillion. The world human population is continuing to rise, while stagnating or falling in most of the OECD. Most importantly, the global middle class could increase to 4.9 billion by 2030, with 85% of the growth coming from Asia (OECD, 2010). With middle-class status comes consumption, but also emissions.

Managing the transition towards a bioeconomy largely hinges on the development of advanced biorefineries (e.g. Iles and Martin, 2013; Kleinschmit et al., 2014). The International Energy Agency (IEA Bioenergy Task 42 Biorefinery, 2012) described a biorefinery as "the sustainable processing of biomass into a spectrum of marketable products (food, feed, materials, chemicals) and energy (fuels, power, heat)". This definition suggests that biorefineries should produce both non-energetic and energetic outputs, and applies to product-driven biorefinery processes. Both primary products and energy-driven processes are considered as true biorefinery approaches provided that sustainable processing of biomass is the final goal (de Jong and Jungmeier, 2015). One of the visions for the bioeconomy is of distributed manufacturing in small- and medium-scale integrated biorefineries. However, this flies in the face of massive fossil fuel and petrochemical economies of scale married to gargantuan subsidies for fossil fuel consumption. Further, this is occurring in a world where an explicit price on carbon and carbon taxation is politically difficult.

For bioeconomy policy makers, the future is complex and multi-faceted. As the first generation of bioeconomy policies comes to a close, the vision of a bioeconomy pitched against grand challenges clearly needs better national and international policies to succeed. This part of the book will address policy issues systematically across global, national and regional scales, and where these intersect and interact. It will use a familiar innovation framework to present these ideas, but will adapt the framework to the specific exigencies of the bioeconomy, illustrated by international examples of policy actions.

The global nature of the societal challenges

In common with bioeconomy goals, the climate agreement reached in Paris in 2015 aims at reducing carbon pollution, while creating more jobs and economic growth driven by low-carbon investments (UNFCCC, 2015). On 5 October 2016, with 97 of the 197 Parties to the Convention having ratified the Paris Agreement, the threshold for entry into force was reached. The agreement subsequently entered into force on 4 November 2016.[1]

At least 97% of actively publishing climate scientists agree that climate-warming trends over the past century are extremely likely due to human activities (Cook et al., 2016).

At the heart of the challenge is the need to decouple economic growth from environmental degradation, and particularly to drastically cut emissions (OECD, 2009). The G7 has called for as-close-as-possible to a 70% reduction on 2010 emissions by 2050 (G7 Germany, 2015). However, when a country doubles its wealth, its emissions rise by about 80% (UNEP, 2010).

At the start of mass production of vehicles, all the major oil reserves remained to be found. At the start of the bioeconomy period, fewer new reserves were being added year-on-year. Conventional oil reserves have been in decline since 1980 (Owen et al., 2010). Discoveries of new oil reserves have dropped to their lowest level in more than 60 years, pointing to potential supply shortages in the next decade (Katakey, 2016). For governments and the private sector alike, resource depletion affects many of the grand challenges. But resource depletion also offers opportunities estimated at USD 80 trillion by 2050 (Cayuela, 2013). By 2100, more than 95% of chemicals and polymers may need to be derived from renewable resources (Devaney et al., 2016).

The relationship between challenges and opportunity is at the heart of replacing the oil barrel and building the bioeconomy. Grand challenges need not be insurmountable obstacles leading to economic despair, but rather the chance to rebuild industry and society in a sustainable manner. Such a process could bring jobs and value added through exploitation of biomass rather than fossil resources. This has been explained as a vision of the future in the United States because "the core petroleum-based feedstock is a limited resource and diversification of feedstocks will provide even greater opportunity for the chemical manufacturing industry" (National Academy of Sciences, 2015).

Past energy and production transitions arguably flourished through "more from more", but the bioeconomy may well have to flourish through "more from less". All bioeconomy aspirations depend on supplies of sustainable biomass (Piotrowski et al., 2015). In the post-fossil fuel world, an increasing proportion of chemicals, plastics, textiles, fuels and electricity will have to come from biomass, which creates greater competition for land (Haberl, 2015). By 2050, the world will need to produce 50-70% more food (FAO, 2009), increasingly under drought conditions (Cook et al., 2015) and on degraded soils (Karlen and Rice, 2015; Nkonya et al., 2016). Herein lies one major conundrum for the bioeconomy – reconciling the conflicting needs of agriculture and industry (Bosch et al., 2015). Inevitably food must come first (e.g. SCAR, 2015; El-Chichakli et al., 2016). The extent to which industrial production can rely on biomass is undetermined (Kim et al., 2011; PBL, 2012).

In another conundrum, bio-based products, including biofuels and bioenergy, are not necessarily sustainable. All biofuels are not equal in this regard, and the same applies to other bio-based products. Evidence is amassing (e.g. Hermann et al., 2007; Weiss et al., 2012; Carus, 2017) that bio-based products can offer environmental advantages, such as significant savings on greenhouse gas (GHG) emissions. However, such benefits cannot be assumed, and products need to be treated case-by-case. Further, estimates of environmental impacts of these products vary greatly, becoming a serious impediment to bio-based production. Critics have raised serious misgivings concerning the use of life cycle analysis (LCA) as the sole tool in environment impact assessment (ANEC, 2012). International standards are required to build the credibility of the industry.

Towards a policy framework for the bioeconomy

Momentum is building across the world towards a policy framework for the bioeconomy. Around 50 countries have adopted the bioeconomy in their economic and innovation strategies. Some have dedicated bioeconomy strategies e.g. Finland, France, Germany, Japan, Malaysia, Norway, South Africa, the United States and the West Nordic countries. Others, such as Austria, Iceland and Tunisia, have plans to develop them. Still others have policies consistent with development of a bioeconomy. These include Australia, Brazil, India, Ireland, Korea, the Netherlands, the People's Republic of China (hereafter "China"), Russian Federation (hereafter "Russia") and Sweden. Bioökonomierat (2015) gives a comprehensive roundup of different national intentions. Countries differ in their priorities, with some focusing more on health and others on bioenergy. Many express the intention to develop a bio-based industry with higher added value products than biofuels or bioenergy.

While national bioeconomy strategies demonstrate intent and commitment, they tend to be short on detail. For this reason, a single document that examines the major policy implications of a bioeconomy, whether a framework is feasible or not, could be useful. Creating such a framework is difficult, however, as the bioeconomy transcends a large range of policy families, including tax, innovation, industry, agriculture, waste and trade. Carus (2014) identified several critical policy areas, many of them under innovation policy. Others can be found in Table 1.1 grouped under three essential categories. These can roughly be translated to supply-side, demand-side and a mixture of both (i.e. cross-cutting measures). This is consistent with the view that both supply- and demand-side policies are needed for effective innovation.

Table 1.1. **Policy inputs for a bioeconomy framework**

Feedstock/technology push	Market pull	Cross-cutting
Local access to feedstocks	Targets and quotas	Standards and norms
International access to feedstocks	Mandates and bans	Certification
R&D subsidy	Public procurement	Skills and education
Pilot and demonstrator support	Labels and raising awareness	Regional clusters
Flagship financial support	Direct financial support for bio-based products	Public acceptance
Tax incentives for industrial R&D	Tax incentives for bio-based products	Knowledge-based capital
Improved investment conditions	Incentives related to GHG emissions (e.g. ETS)	
Technology clusters	Taxes on fossil carbon	
Governance and regulation	Removing fossil fuel subsidies	

Note: R&D = research and development; GHG = greenhouse gas; ETS = emissions trading system.

Source: Adapted from Carus (2014), "Strategy for a rethink of the policy framework for the bio-based economy".

Demand is a major potential source of innovation, yet government policy may not recognise it as such (Edler and Georghiou, 2007). Historically, OECD countries have tended to rely on macroeconomic policies (e.g. monetary and fiscal) and framework conditions (e.g. competition, tax or entrepreneurship policy) to support market demand and avoid distortion. In recent years, however, OECD countries and emerging economies such as Brazil and China have used more targeted demand-side innovation policies. These include measures such as public procurement, regulation, standards, consumer policies and user-led innovation initiatives. They also include lead market initiatives to address market and system failures in areas with pressing social needs (OECD, 2011).

Experience in OECD countries has shown that use of such demand-side policies remains limited to areas in which the market alone cannot meet societal needs (e.g. environment) or in which private and public markets intersect (e.g. energy supply). Both the environment and energy drive bioeconomy policy goals. This focus on the demand side also reflects a general perception that traditional supply-side policies – despite design refinements over recent decades – have not brought innovation performance and productivity to desired levels.

Policy at multiple scales

The complexity of bioeconomy policy is partly due to the multiple scales of action required (Figure 1.2). These scales range from regional development (e.g. biorefinery deployment) through to national research and development (R&D) into synthetic biology, information technology (IT) convergence and automation to global issues of biomass and its sustainability. The distributed bioeconomy manufacturing model calls for a "glocal" approach i.e. both global and local. It stresses the importance of locating the growing industry close to both raw materials and the goods and energy that are produced and consumed (McCormick and Kautto, 2013). Unlike the petrochemicals model, the success of the bioeconomy manufacturing model does not rely on economies of scale. This could prove to be a major challenge (IHS Markit, 2015).

Figure 1.2. **Bioeconomy policy moves from regional to global**

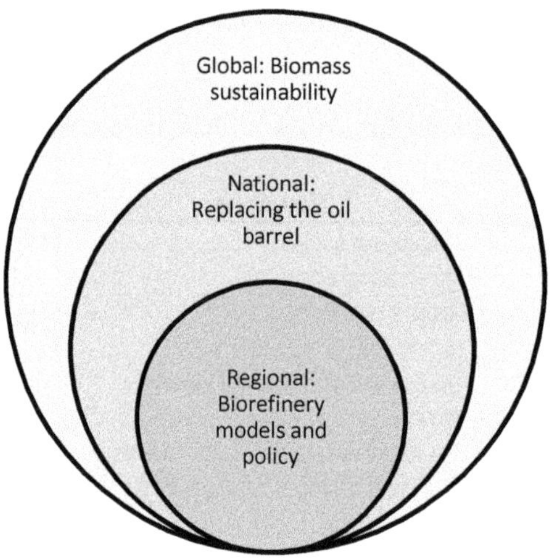

Source: Philp (2018), "The bioeconomy, the challenge of the century for policy makers", http://dx.doi.org/10.1016/j.nbt.2017.04.004.

The bioeconomy arguably has regions at its heart; building future production facilities in regions throws up both threats and opportunities. However, a large amount of R&D is still required across a wide range of topics. This speaks more to national-level funding, especially as biotechnology depends highly on basic science. Further, prevention of over-exploitation of natural resources is a matter for global effort. Treating policy in these separate, but related scales, hopefully removes confusion and points it more directly to where specific measures are needed.

The book may seem to paint a rosy picture of international co-operation with plenty of infrastructure investment and therefore a booming bioeconomy sector. In fact, it demonstrates the beginning of the transition to a new model of production based on decentralisation and sustainability (Il Bioeconomista, 2016). Several countries are strong in bioeconomy research and relatively poor in deployment. In terms of biorefining capacity, perhaps Finland is in the lead. Great hopes are pinned on the cellulosic biorefineries, but they are worryingly susceptible to technical failure. To date, cellulosic ethanol volumes are still but a trickle and depend on government largesse (Peplow, 2014). Clearly, research progress is way ahead of full-scale deployment, which is not surprising in such a young industry. This book points to the major policy needed to redress the balance between R&D and commercial success – a long and tortuous journey.

Schieb et al. (2015) suggest the need to increase biorefineries to 300-400, both in the United States and Europe, for the industrial bioeconomy to succeed. That represents a very large investment, most of which will need to come from the private sector. The bio-based private sector, however, needs stable and long-term policies to invest in risky projects. Thus governments need to share the same view of the future of the bioeconomy. A view from Australia could be easily generalised to any country: future prospects for industrial biotechnology are "predicated on governments taking a long view of the nation's future strategic position in an industrial world that will be green of necessity" (Glenn, 2017).

Despite its growing pains, the bioeconomy is marching forward. Il Bioeconomista (2015) suggests providing the bioeconomy with a 15-25 year competitive advantage over the fossil industry. At first glance, this seems an expensive option. However, after a century of operation, fossil industries enjoy astronomically high subsidies. Further, even the fossil industry has accepted the reality of climate change, and recognises the need to adapt. Progress is being made when the Rockefeller Family Fund trustees say: "While the global community works to eliminate the use of fossil fuels, it makes little sense – financially or ethically – to continue holding investments in these companies" (Cunningham, 2016). Even Saudi Arabia plans to diversify its economy and end its reliance on oil in the near future (Kingdom of Saudi Arabia, 2017).

Note

1. http://unfccc.int/paris_agreement/items/9444.php.

References

ANEC (2012), "ANEC position. Environmental assessment goes astray. A critique of environmental footprint methodology and its ingredients", ANEC-ENV-2012-G-008final, European Association for the Co-ordination of Consumer Representation in Standardisation, Brussels.

Bennett, S.J. and P.J.G. Pearson (2009), "From petrochemical complexes to biorefineries? The past and prospective co-evolution of liquid fuels and chemicals production in the UK", *Chemical Engineering Research and Design*, Vol. 87/9, Elsevier, Amsterdam, pp. 1120-1139.

Bioökonomierat (2018), "Internationale Bioökonomiestrategien", webpage, http://biooekonomierat.de/biooekonomie/international/ (accessed 30 January 2018).

Bioökonomierat (2015), "Bioeconomy policy (part II): Synopsis of national strategies around the world", Bioökonomierat, Berlin.

Bosch, R. et al. (29 July 2015), "Define biomass sustainability", Nature blog, Vol. 523, Nature Publishing Group, London, pp. 526-527.

Carus, M. (2017), "Bio-based economy and climate change – Important links, pitfalls and opportunities. Nova-Institute", *Industrial Biotechnology*, April, Vol. 13/2, Mary Ann Liebert, New Rochelle, United States, pp. 41-51.

Carus, M. (2014), "Strategy for a rethink of the policy framework for the bio-based economy", presentation at the OECD workshop on bio-based production", 9 October, Turin.

Cayuela, R. (2013), *The Future of the Chemical Industry by 2050*, Wiley-VCH, Weinheim.

Cook, B.I. et al. (2015), "Unprecedented 21st century drought risk in the American Southwest and Central Plains", *Science Advances,* Vol. 1/1, American Association for the Advancement of Science, New York, e1400082, http://dx.doi.org/10.1126/sciadv.1400082.

Cook, J. et al. (2016), "Consensus on consensus: A synthesis of consensus estimates on human-caused global warming", *Environmental Research Letters*, Vol. 11/4, IOPscience, Bristol, http://dx.doi.org/10.1088/1748-9326/11/4/048002.

Cunningham, N. (24 March 2016), "Rockefeller family fund blasts ExxonMobil, pledges divestment from fossil fuels", Oil Price blog, http://oilprice.com/Energy/Energy-General/Rockefeller-Family-Fund-Blasts-ExxonMobil-Pledges-Divestment-From-Fossil-Fuels.html.

de Jong, E. and G. Jungmeier (2015), "Biorefinery concepts in comparison to petrochemical refineries", in *Industrial Biorefineries and White Biotechnology*, Elsevier, Amsterdam, http://dx.doi.org/10.1016/B978-0-444-63453-5.00001-X.

Devaney, L. et al. (2016), "Imagine the future", *Taste of Science*, 29 May, Wageningen, The Netherlands, www.tasteofscience.com/articles/609/imagine-the-future.html.

Edler, J. and L. Georghiou (2007), "Public procurement and innovation – Resurrecting the demand side", *Research Policy*, Vol. 36, Elsevier, Amsterdam, pp. 949-963.

El-Chichakli, B., et al. (2016), "Five cornerstones of a global bioeconomy", *Nature*, Vol. 535/7611, Nature Publishing Group, London, pp. 221-223.

FAO (2009), *The State of Food and Agriculture: Livestock in the Balance*, Food and Agriculture Organization of the United Nations, Rome.

G7 Germany (2015), "Think ahead. Act together", Leaders' Declaration G7 Summit 7-8 June, https://sustainabledevelopment.un.org/content/documents/7320LEADERS%20STATEMENT_FINAL_CLEAN.pdf.

Glenn, D. (2017), "The long view: Policy directions in Australia to support the industrial biotechnology sector", *Industrial Biotechnology*, Vol. 13/2, Mary Ann Liebert, New Rochelle, United States, pp. 72-75.

Haberl, H. (2015), "Competition for land: A sociometabolic perspective", *Ecological Economics*, Vol. 119, Elsevier, Amsterdam, pp. 424-431.

Hermann, B.G. et al. (2007), "Producing bio-based bulk chemicals using industrial biotechnology saves energy and combats climate change", *Environmental Science and Technology*, Vol. 41/22, *Science*, American Association for the Advancement of Science, Washington, DC.

IEA Bioenergy Task 42 Biorefinery (2012), "Bio-based chemicals. Value added products from biorefineries", International Energy Agency, Paris, www.ieabioenergy.com/publications/bio-based-chemicals-value-added-products-from-biorefineries.

IHS Markit (2015), *Chemical Economics Handbook: Petrochemical Industry Overview*, April, IHS Markit, London.

Il Bioeconomista (30 September 2016), "Novamont opens world's first commercial-scale bio-BDO plant in North-eastern Italy", Il Bioeconomista blog, https://ilbioeconomista.com/2016/09/30/novamont-opens-worlds-first-commercial-scale-bio-bdo-plant-in-north-eastern-italy.

Il Bioeconomista (12 October 2015), "Ten billion euros of investment in advanced biofuels", Il Bioeconomista blog, https://ilbioeconomista.com/2015/10/12/ten-billion-euros-of-investment-in-advanced-biofuels.

Iles, A. and A.N. Martin (2013), "Expanding bioplastics production: Sustainable business innovation in the chemical industry", *Journal of Cleaner Production*, Vol. 45, Elsevier, Amsterdam, pp. 38-49.

Karlen, D.L. and C.W. Rice (2015), "Soil degradation: Will humankind ever learn?", *Sustainability*, Vol. 7/9, MPDI, Basel, pp. 12490-12501, http://dx.doi.org/10.3390/su70912490.

Katakey, R. (2016), "Oil discoveries have shrunk to a six-decade low", Bloomberg News, 23 May, www.bloomberg.com/news/articles/2016-05-23/oil-discoveries-slumped-to-lowest-since-1952-as-budgets-slashed.

Kim, J. et al. (2011), "Optimal design and global sensitivity analysis of biomass supply chain networks for biofuels under uncertainty", *Computers & Chemical Engineering*, Vol. 35/9, Elsevier, Amsterdam, pp. 1738-1751.

Kingdom of Saudi Arabia (2017), *Saudi Vision 2030,* Media Center website, http://vision2030.gov.sa/en/media-center (accessed 7 December 2017).

Kleinschmit, D. et al. (2014), "Shades of green: A social scientific view on bioeconomy in the forest sector", *Scandinavian Journal of Forestry Research*, Vol. 29/4, Taylor & Francis Online, pp. 402-410.

McCormick, K. and N. Kautto (2013), "The bioeconomy in Europe: An overview", *Sustainability*, Vol. 5/6, MPDI, Basel, pp. 2589-2608.

National Academy of Sciences (2015), *Industrialization of Biology: A Roadmap to Accelerate the Advanced Manufacturing of Chemicals*, National Academy of Sciences, Washington, DC.

Nkonya, E. et al. (eds.) (2016), *Economics of Land Degradation and Improvement – A Global Assessment for Sustainable Development*, Springer, Cham, Switzerland.

OECD (2017), *The Next Production Revolution: Implications for Governments and Business*, OECD Publishing, Paris, http://dx.doi.org/10.1787/9789264271036-en.

OECD (2011), *Demand-side Innovation Policies*, OECD Publishing, Paris, http://dx.doi.org/10.1787/9789264098886-en.

OECD (2010), "The emerging middle class in developing countries", *OECD Development Centre Working Papers*, No. 285, OECD Publishing, Paris, http://dx.doi.org/10.1787/5kmmp8lncrns-en.

OECD (2009), *The Bioeconomy to 2030: Designing a Policy Agenda*, OECD Publishing, Paris, http://dx.doi.org/10.1787/9789264056886-en.

Owen, N.A. et al. (2010), "The status of conventional world oil reserves – Hype or cause for concern?", *Energy Policy*, Vol. 38, Elsevier, Amsterdam, pp. 4743-4749.

PBL (2012), "Sustainability of biomass in a bio-based economy: A quick-scan analysis of the biomass demand of a bio-based economy in 2030 compared to the sustainable supply", *PBL Note*, No. 500143001, Netherlands Environmental Assessment Agency, Amsterdam.

Peplow, M. (11 March 2014), "Cellulosic ethanol fights for life. Pioneering biofuel producers hope that US government largesse will ease their way into a tough market", Nature blog, Vol. 507/7491, www.nature.com/news/cellulosic-ethanol-fights-for-life-1.14856.

Philp, J. (2018), "The bioeconomy, the challenge of the century for policy makers", *New Biotechnology*, Vol. 40 Part A, Elsevier, Amsterdam, pp. 11-19, http://dx.doi.org/10.1016/j.nbt.2017.04.004.

Piotrowski, S. et al. (2015), "Global bioeconomy in the conflict between biomass supply and demand", *nova-Paper on bio-based economy*, No. 7, Nova Institute, Hürth, Germany.

SCAR (2015), "Sustainable agriculture, forestry and fisheries in the bioeconomy – a challenge for Europe", Standing Committee on Agricultural Research 4th Foresight Exercise, 15 June, European Commission, Brussels.

Schieb, P.-A. et al. (2015), *Biorefinery 2030: Future Prospects for the Bioeconomy*, Springer-Verlag, Berlin.

Schiermeier, Q. (10 April 2013), "Renewable power: Germany's energy gamble", Nature blog, Vol. 496, Nature Publishing Group, London, pp. 156-158.

UNEP (2010), *Assessing the Environmental Impacts of Consumption and Production: Priority Products and Materials*, United Nations Environment Programme, Washington, DC.

UNFCCC (2015), "Adoption of the Paris agreement", FCCC/CP/2015/L.9/Rev.1, United Nations, New York, http://unfccc.int/resource/docs/2015/cop21/eng/l09r01.pdf.

Weiss M., et al. (2012), "A review of the environmental impacts of bio-based materials", *Journal of Industrial Ecology*, Vol. 16, Wiley Online Library, pp. S169-S181.

Part I.

Biomass sustainability

Chapter 2.

Reconciling food and industrial needs for biomass

This chapter is mainly about non-OECD countries and their developing bioeconomies. They are central to the development of a globalised bioeconomy and to the sustainable future of OECD countries' bioeconomies. Future projections see the need for significantly increased agricultural output to feed a growing population. And yet there seems to be limited capacity for increased land use (extensification). This dilemma could bring OECD countries and partner economies into competing use for biomass. Many OECD countries will be net biomass importers, while many developing and poorer nations can be expected to be exporters of biomass. Nations could easily collide with each other through biomass disputes. A top priority for policy makers is to reconcile the food and industrial demands of biomass to prevent negative effects in some nations being created through positive effects in others. A sustainable bioeconomy cannot be produced through such poorly distributed benefits.

Introduction

In the post-fossil world, major sources of carbon will still be required. The internal combustion engine is ultimately replaceable, but society will forever need chemicals to maintain the lifestyle of developed countries, and to bring this more comfortable lifestyle to other countries. The foreseeable sources of carbon are renewable biomass carbon and waste industrial gases. As the latter is the target of climate change mitigation and waste reduction policies, it also will dwindle with time. Therefore, renewable biomass carbon is envisaged to become a major source of carbon for chemicals, plastics, textiles, materials and aviation fuels of the future.

This immediately creates a dilemma as the food and industrial uses of biomass clearly come into competition. While there are fewer people hungry than ever before, food security is still elusive in many countries (FAO, 2017). Moreover, this conflicting use of biomass has geographical and geopolitical implications. On the one hand, many OECD countries can be expected to be importers of biomass (some already are) due to a shortage of land and high population densities. On the other, many partner economies are rich in biomass, including Brazil, India, Indonesia, Malaysia, the People's Republic of China (hereafter "China") and the Russian Federation. The latter may be tempted simply to export natural resources, as was sometimes the case in the past.

Two problems with an export-focused strategy for developing nations would be:

- Simply exporting natural resources may inhibit technological development. There is far greater value added for a nation to develop the technologies of a bioeconomy (e.g. industrial biotechnology, green chemistry and modern agricultural practices).
- In the absence of strong governance, biomass could be over-exploited as a resource, resulting in market and societal failures such as deforestation and soil destruction. Many potential social risks can be imagined (e.g. warlordism, displacement of landowners, threats to traditional lifestyles, and job losses and gains within the same society) (Obidzinski et al., 2012).

Biomass flows

Global biomass flows point to a trade issue for OECD countries, many of which are advanced economies that lack access to large amounts of biomass within their boundaries. Therefore, bioeconomies in these nations rely on biomass imports. This may encourage exporting countries to grow and harvest biomass unsustainably. If this involved food crops destined for industrial use, it could be a potential threat to food security. There is a massive quandary at the beating heart of the bio-production concept – how to reconcile the food and feed use of biomass with the needs of industrial production.

Figure 2.1 clearly shows that biomass flows should concern the OECD: every single arrow-head points to an OECD country or region. Further, there is a significant convergence on Western Europe.

The flow of world biomass shipping routes towards the OECD presents risks for both parties. OECD countries that lack biomass resources may switch their dependence on oil exporters to biomass exporters. Consequently, bio-production may fail to achieve policy goals like energy security. Biomass exporting countries that rely simply on exporting raw materials would miss the opportunity to create the greater value-added bio-production industries. This, in turn, could lead to unsustainable practices, particularly over-exploitation.

Figure 2.1. **Major world biomass shipping routes in 2011**

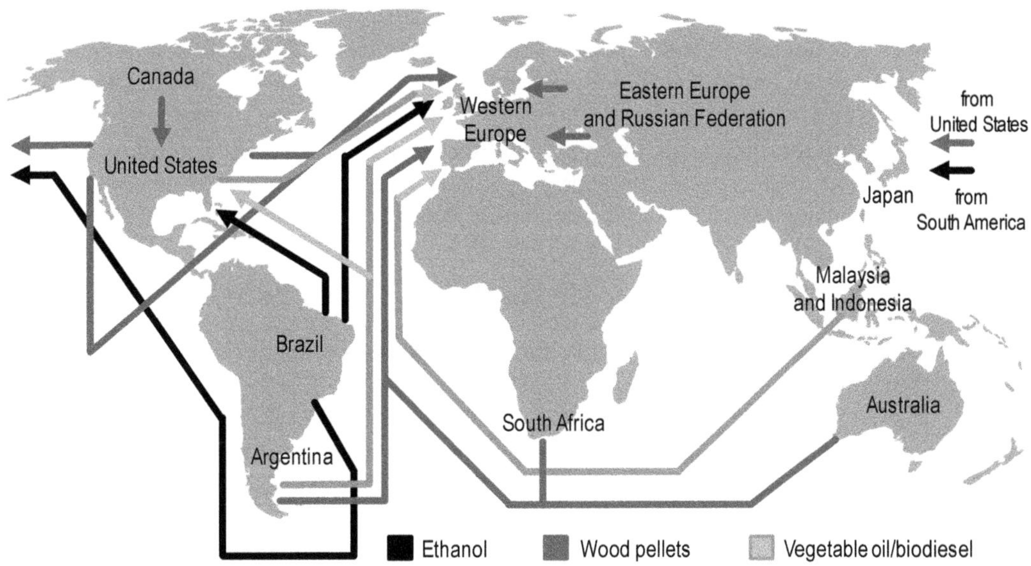

Source: Redrawn from BP-EBI (2014), *Biomass in the Energy Industry, An Introduction.*

The bioeconomy can deliver great benefits for society as a whole in terms of both energy and food security. For example, it could distribute energy resources more evenly rather than concentrate them in small, politically unstable regions of the world. Further, agricultural productivity (the value added per agricultural worker) of many Asian countries is much lower than in developed countries (World Bank, 2017). Farming is characterised by small farms, subsistence farming and high levels of poverty. Changing both agricultural practices and the application of modern biotechnology could thus greatly enhance food supply.

Box 2.1. **Controlling deforestation in Liberia**

Forest covers more than 40% of Liberia, a country considered one of West Africa's most important carbon sinks and biodiversity hotspots. The United Nations estimates that 30 000 hectares of primary Liberian forest is cleared each year. The country's administration, backed by more than USD 150 million of international aid, is driving policy aimed at enabling the country and communities to make money from reduced carbon emissions. First, carbon levels are measured in a forest. Then, if the land is not cleared, the carbon that is retained in the forest – or not emitted through clearing – can be sold as offsets.

Norway is providing USD 70 million to help Liberia develop the policy framework and create capacity to implement the changes. It is providing a further USD 80 million to pay for the first carbon offsets. Other governments and private investors are welcome to buy the offsets. It will take time to see whether such a system could succeed, but this could be a test-bed for deforestation prevention. A bioeconomy is likely to stimulate markets for wood further; this policy is consistent with reducing emissions, one of the major policy goals of a bioeconomy.

Source: Aglionby (8 April 2016), "Green revolution aims to stem deforestation in Liberia", www.ft.com/intl/cms/s/0/9e596f2e-dbbb-11e5-a72f-1e7744c66818.html#axzz45DdIxzJY.

A large reliance on forestry for industrial biomass could lead to environmental degradation as a result of the direct consequences of deforestation. Logging in the past has created conflict, including violence. Illegal logging is already costing nations tens of billions of dollars each year, and tropical deforestation contributes 12% of total anthropogenic carbon dioxide emissions globally (Lynch et al., 2013). Therefore, illegal logging works against two founding policy goals of a bioeconomy – economic growth and climate change mitigation. Paying to prevent deforestation is likely to be contentious, but contributions from OECD countries may be less expensive than letting it continue unabated (Box 2.1).

The twin dilemmas of food and energy security are intimately linked

The case of India shows how easily the bioeconomy could develop unevenly. Like most countries, India imports crude oil at great expense. During the next 25 years, demand for electricity in India is expected to increase five-fold. Many believe the biotechnology sector could help solve India's growing energy problem and its need for energy security. India faces the ultimate dilemma of the bioeconomy: can it produce sufficient biomass to contribute to energy security and economic growth through bio-based production, while still feeding the nation? Many nations with bioeconomy aspirations face the same dilemma. Korea imports 97% of its energy, which still comes from fossil fuel reserves. Many countries in Africa are in the same position, if not worse, as their economies are developing more slowly than some in Southeast Asia. The Japanese government projects that the population of Japan will seriously decline by 2050. Moreover, Japan has a dwindling number of farmers, who are ageing; the average age of Japanese farmers was 65.9 years in 2011. They are also farming very small plots. This poses problems for agricultural vitality (Karan, 2005). Farmers' children do not want to stay in farming. This is by no means unique to Japan. In China, for example, the rural population is also declining, the average age of farmers is rising and fewer young people are choosing to farm as a vocation (Yang, 2013).

Land potential: Tension between food and industrial use of biomass

Table 2.1 contains data that highlight the tensions between food and industrial use of biomass. It demonstrates that if OECD countries become active in world food security, there will soon be no farmland left for industrial use.

Table 2.1. **Land potentials (farmland) for non-food use, "business as usual" scenario**

	2010	2015	2020	2050	2010	2015	2020	2050
Europe	102 717	115 134	127 096	171 446	44 531	20 315	0	0
North America	65 621	59 090	53 709	33 144	27 759	10 135	0	0
Central America	-3 545	-11 765	18 639	-42 219	1 171	446	0	0
South America	35 786	29 132	24 170	18 066	21 182	9 364	0	0
Oceania	33 157	28 185	23 362	-6 416	14 026	4 834	0	0
Asia	-62 219	-113 430	-153 786	-292 920	18 540	6 734	0	0
Africa	-56 818	-91 310	-121 677	-322 022	6 385	3 717	0	0

Note: Figures are x 1 000 hectares.

Source: Adapted from DBFZ (2011), *Global and Regional Spatial Distribution of Biomass Potentials.*

This table shows the farmland potentials (i.e. farmland for non-food use) in the "business as usual" (BAU) scenario developed by the authors. The left side of the table represents farmland potential if the countries in these continents (134 countries in total) do not take part in food security for nations in food deficit. The right side represents the remaining non-food land potential when the same group of countries participates on a *pro-rata* basis in exports to cover the deficit food supply of the import countries. In the following quote, bold text is the author's emphasis.

> The data for the "BAU" scenario indicates that no more farmland potentials for non-food use will be globally available from the year 2020. However, there is still grassland for non-food use. Since no more farmland is available for non-food purposes, the big surplus states for agricultural primary products, such as Europe, North America and South America would have to export as of 2020 all agricultural primary products, which are no longer needed for their own food supply, into countries in deficit (mainly Asia and Africa). (DBFZ, 2011.)

If this analysis is correct, and accepted worldwide, then using waste sources for biorefining is not a luxury, but an absolute necessity. What is more, this is a near-term situation. However, the figures may vary according to assumptions and this table relates to only one resource (farmland). The overall discussion considers more resources, including forests, residual biomass, the marine environment and waste gases. It is this variability in assumptions that leads to such great variety in studies.

Conclusions

Far greater attention must be paid to the conundrum of food-industrial use of biomass. The OECD could host a future event, but should include more stakeholders such as the Food and Agriculture Organization of the United Nations (FAO), the European Commission and the German Bioeconomy Council. Other key players are sustainability certification organisations such as International Sustainability and Carbon Certification, and the Global Bioenergy Partnership. Such an event could also engage government ministries with direct interest such as the departments of energy and agriculture in the United States.

There are countries that are biomass-poor and those that are biomass-rich. A sustainable bioeconomy would have a better balance of power among nations, but this requires serious consideration of another critical balance – between food and industrial use of biomass. An international trade of biomass that achieves industry security in consumer nations (many OECD countries), but food insecurity in exporting nations would defeat the purpose of a bioeconomy. While it might achieve climate change mitigation, it may also threaten food and energy security.

The policy regime governing this transition is enormously complex. Clearly, international and domestic policy will be equally essential. There is much to be gained in a bioeconomy. However, if it does not achieve sustainability, then it will miss a great opportunity for inclusive economic growth.

References

Aglionby, J. (8 April 2016), "Green revolution aims to stem deforestation in Liberia", Financial Times blog, www.ft.com/intl/cms/s/0/9e596f2e-dbbb-11e5-a72f-1e7744c66818.html#axzz45DdIxzJY.

BP-EBI (2014), *Biomass in the Energy Industry, An Introduction*, BP, Energy Biosciences Institute, London.

DBFZ (2011), *Global and Regional Spatial Distribution of Biomass Potentials*, Deutsches Biomasse Forschungszentrum, Leipzig.

FAO (2017), "Food Security Statistics", webpage, www.fao.org/economic/ess/ess-fs/en (accessed 7 December 2017).

Karan, P.P. (2005), *Japan in the Twenty-first Century: Environment, Economy and Society*, University Press of Kentucky, Lexington.

Lynch, J. et al. (2013), "Choose satellites to monitor deforestation", *Nature*, Vol. 496, Nature Publishing Group, London, pp. 293-294.

Obidzinski, K. et al. (2012), "Environmental and social impacts of oil palm plantations and their implications for biofuel production in Indonesia", *Ecology and Society*, Vol. 17/1, Acadia University, Wolfville, Canada, pp. 25, http://dx.doi.org/10.5751/ES-04775-170125.

World Bank (2017), Agricultural output and productivity, 2000-15, in *World Development Indicators*, http://wdi.worldbank.org/table/3.3.

Yang, Z. (2013), "Demographic changes in China's farmers: The future of farming in China", *Asian Social Science*, Vol. 9/7, Canadian Center for Scientific Education, North York, pp. 136-143.

Chapter 3.

Measuring biomass potential and sustainability

This chapter examines the issues around setting biomass sustainability as an essential element to a future bioeconomy. Use of biomass for bio-based production in ambitious bioeconomy plans is fraught with the risk of unsustainable, over-exploitation of natural resources. Developing only modest bioeconomy strategies is one option, but may not achieve the longer-term goals of highly ambitious reductions of greenhouse gas (GHG) emissions. Another option is to create ambitious bioeconomy plans that make biomass production and use more efficient. However, studies also point out that more land is needed to produce biomass. So a dual strategy can be envisioned – land intensification and extensification. Each brings its own problems; the most frequently discussed relate to sustainability, and the inevitable competition for land between food and industrial use. There is no international agreement yet on how to measure biomass sustainability. As a result, estimates of biomass potential (how much can be grown sustainably) vary greatly. New institutions may be necessary to harmonise sustainability assessments.

Introduction

Biomass potential refers to how much biomass can actually be grown at any scale – regional, national, supranational or global. Measurements generally fall into three different categories – agricultural, forestry and waste biomass – and may or may not consider marine biomass as the studies; they are usually focused on the sustainability of terrestrial sources. However, marine biomass will play important roles in securing biomass in the future. As seen in a later part of this book, marine biorefining models are among the least developed for mainly technical reasons.

Future bioeconomy policy must also consider the roles of non-biomass carbon that exist in huge quantities but are as yet barely used. These can take pressure off land use for industrial sources of biomass, allaying fears about using biomass for industry when the top priority is for food. Industrial sources of CO_2 are already used for specific purposes, such as for carbonating soft drinks. However, this hardly scratches the surface of the potential of waste CO_2 and other industrial gases for biorefining such as CO and H_2. The use of these waste gases in fermentation has already begun, but the technologies are in their infancy. A strong focus of biomass sustainability thinking and policy is how these vast reserves of carbon could, in future, greatly alleviate pressure on land.

Sustainable biomass potential can be defined as the fraction of the technical biomass potential that does not oppose the general principles of sustainable development, i.e. the fraction that can be exploited in an economically viable manner without causing social or ecological damage (Rettenmaier, 2008).

How much biomass can be grown and how much is needed: Biomass potential

The recurring theme around biomass potential is "uncertainty". There are no internationally accepted metrics or tools to apply questions of sustainability to biomass (Bosch et al., 2015). Not surprisingly, biomass potential estimates are extremely variable. Working from 17 separate studies, Saygin et al. (2014) identified a discrepancy in estimates of biomass potential of 20-fold from highest to lowest (75 to as high as 1 500 exajoules per year [EJ/yr] in 2050). Figure 3.1 helps illustrate these discrepancies. Schueler et al. (2016) observed a range of technically available potentials between 50-500 EJ per year by mid-century. Applying sustainability criteria to the available biomass potential decreases it considerably.

Types of biomass potential assessment

Several studies over the past years have used a range of techniques to estimate the available land for bioenergy production – from simple data assumptions to robust high-resolution land mapping. Hence, large differences in estimates exist. Most studies provide detailed insights into future biomass potential, but fail to include all critical factors involved in the assessment. An "ideal" study to evaluate biomass potential should consider global and regional trends, as well as local conditions such as soil types, water availability, possibility of irrigation and land-use planning. It should further consider biodiversity and soil quality (Dornburg et al., 2008). However, this ideal may only be possible at a restricted regional level, if at all. These crucial factors can hugely alter the range of sustainable biomass potential. Seidenberger et al. (2008) have attempted to compile global biomass potential ranges from 18 different studies (Figure 3.1).

Figure 3.1. **A compilation of estimates for global biomass potentials**

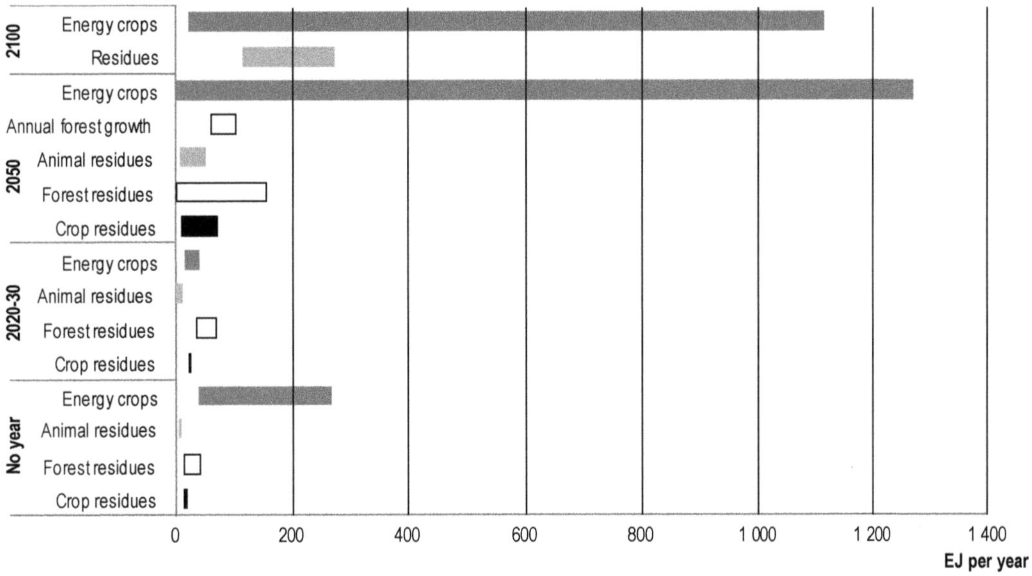

Note: EJ = exajoules.

Source: Adapted from Seidenberger et al. (2008), *Global Biomass Potentials – Investigation and Assessment of Data, Remote Sensing in Biomass Potential Research, and Country-specific Energy Crop Potentials.*

This shows the minimum and maximum potentials estimated by different studies.

Discrepancies in biomass potential estimates

Studies attempting to estimate the availability of biomass have considered both optimistic and pessimistic approaches. The range varies for several reasons. There are different objectives elaborated over different time frames. Many biomass potential studies have future estimates until 2050, but less information is available on the short term. Various methodologies and approaches have also been used to estimate biomass potential. In addition, the lack of a commonly agreed definition on the types of biomass (forest residues, harvest and process residues) influences estimates. This leads to different data sets generated with different criteria. Estimates depend on developing scenarios, but scenario assumptions vary widely. Further, some studies lack transparency and may omit factors. Finally, the geographical scope of different studies can make results confusing to compare.

Calculating biomass potential and estimating the size of a potential bioeconomy

Numerous options exist for replacing liquid fossil fuels in the long term. Material uses, for example, include plastics, chemicals and textiles. But once the options are examined, the only serious contender in terms of quantity is biomass. Bioenergy is the most important renewable energy option, at present and in the medium term (Ladanai and Vinterbäck, 2009). However, bioenergy also offers the greatest potential for unsustainable, over-use of biomass due to the volumes required.

Dual use of biomass is effectively a competition for land with food use always taking first priority. The availability of sustainable biomass as a future substitute for fossil resources depends on the available land for biomass cultivation, and options to use the biomass produced in agriculture and forestry more efficiently.

To understand these two factors, it is necessary to know how biomass flows in agriculture and forestry. If these flows of biomass over the world can be quantified, the potential to use more biomass for new applications without disturbing current applications can be assessed. Unsurprisingly, there have been many estimates of biomass flows, all with high levels of uncertainty.

One main source of uncertainty is the underlying assumption regarding the amount of unused agricultural land available for cultivation of bioenergy crops, and to what extent natural grasslands contribute to this potential. In particular, assumptions regarding future agricultural productivity and future consumption of animal products have a great impact on the results. Furthermore, the amounts of available waste and residue resources strongly depend on the still uncertain future demand for other applications such as animal feed and soil quality improvers. Moreover, estimates are necessarily indicative because future trade is uncertain.

The energy content of agricultural crops, including their residues produced across the world, is estimated at 200 EJ; grass- and rangelands produce about 115 EJ. Both mainly deliver the inputs for human food. Most of the energy is not available for the energy system because it is vital in the livestock system and also for people. Setting aside the unused and sometimes burned crop residues for energy could increase the extraction by about 24 EJ. This assumption considers sustainable soil carbon management (roughly half of the above-ground carbon should remain in the soil). Other potential energy sources are better use of waste flows from industrial processing and consumption. This could produce an additional 21 EJ.

Another uncertainty around estimating biomass potential is the future extent of the bioeconomy, which is decided politically as well as scientifically. Estimates therefore often rely on scenario development. The Netherlands Environmental Assessment Agency (PBL) defined three different scenarios for biomass potential to 2050 (PBL, 2012):

- High:
  - very productive agriculture, leaving land for energy crops
  - use of almost all sustainably available residues and waste
  - successful new developments.
- Mid:
  - more productive agriculture, but quite limited land for energy
  - use of about half of the sustainably available residues and waste
  - only a few new developments for niche markets.
- Low:
  - unsustainable land use for energy crops
  - use of only a small part of residues and wastes
  - no new developments.

These scenarios assessed the global biomass potential for energy use in 2050. According to the PBL and ECN (Energy Research Centre of the Netherlands), most studies estimate the potential availability of 150-250 EJ of sustainable biomass by 2050, which is considered economically feasible. For 2030, PBL considers 100 EJ as a "realistic" estimate and 200 EJ as an "optimistic" estimate of available sustainable biomass on the world market.

As a further illustration, the potential for Europe (including trade) would be about 10 EJ based on "mid" expectations assuming an equal distribution per capita in 2050. With a distribution based on income, the potential might double. The European Union will therefore probably depend on the world market to supply biomass for its bioeconomy in the future.

Policy implications

- The total supply of sustainable biomass in 2030 may be enough to fulfil the demand in a 10% bio-based economy (PBL, 2012).
- A highly ambitious bioeconomy increases the risk of a non-sustainable supply and over-exploitation of natural resources.
- In light of growing trade, the numbers need to be continuously re-assessed.
- Looking beyond 2030 to 2050, many new initiatives and technologies will be required to reach the potential of sustainable biomass.
- Algal biomass may be useful in the future, but costs are currently much too high for bioenergy. However, if future development of aquatic biomass is successful, this type of biomass production could offer new possibilities. Suggesting any number for future potential is just a first guess. At this stage, feasibility studies and research and development (R&D) support are the most obvious policy options.

Experience with estimating biomass potential in the United States

Over the past decade, the United States has made a concerted effort to discover the national biomass potential. This resulted in the first *Billion Ton Report*, completed in 2005 and subsequently updated (US DOE, 2017, 2016, 2011, 2005). The basics remain the same throughout these reports: the United States, depending on assumptions, might produce 1 billion tonnes of dry biomass per annum. This would substitute 30% of gasoline requirements with renewable biofuels. The authors estimate the country uses 365 million dry tonnes of agricultural crops, forestry resources and waste to generate biofuels, renewable chemicals and other bio-based materials. The most recent updates also evaluate the policies and economic conditions needed to direct investment to the bio-based economy and to build the biorefineries that will use potential biomass resources.

Like other biomass potential studies, the *Billion Ton* studies are based upon scenarios:

- *The baseline scenario*: Combined resources from forests and agricultural lands total about 473 million dry tonnes at USD 60 per dry tonne or less (about 45% is used and the remainder is potential additional biomass). By 2030, estimated resources increase to nearly 1.1 billion dry tonnes (about 30% would be projected as already used biomass and 70% as potentially additional).
- *The high-yield scenario*: Total resource ranges from nearly 1.4 billion to over 1.6 billion dry tonnes annually of which 80% is potentially additional biomass. No high-yield scenario was evaluated for forest resources, except for woody crops.

The adopted methodology allowed estimates of the biomass potential of different sub-sectors (Figure 3.2). This is critical when developing future scenarios. Technological developments in one sub-sector may offset a lack of development in another. And as they change, this may alert governments to future policy needs.

Figure 3.2. **Biomass potential in the baseline scenario (a) and high-yield scenario (b)**

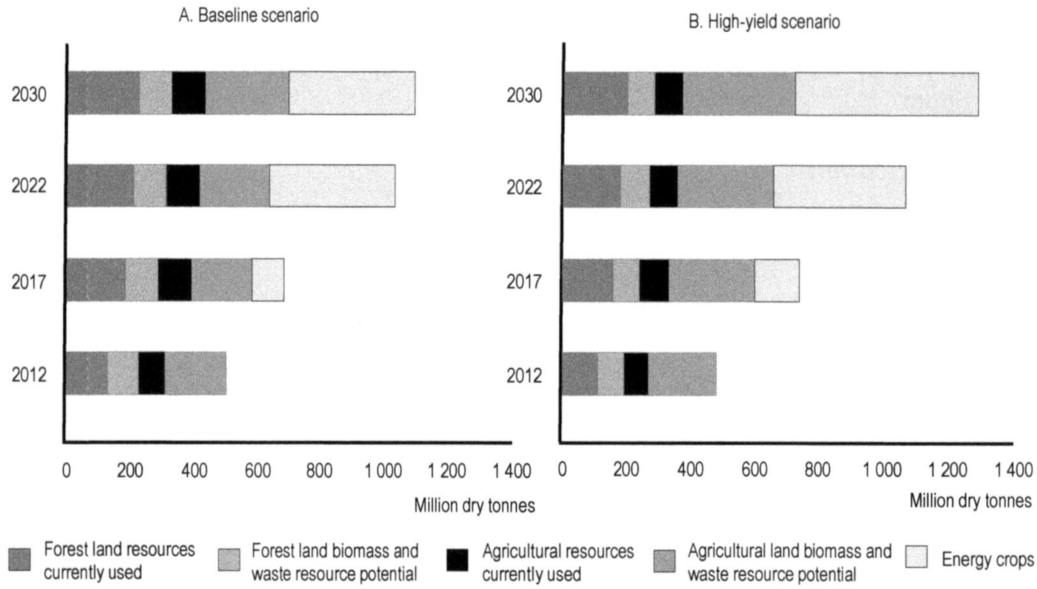

Source: Stokes (2014), "The 'Billion Ton Update': Methodologies and implications".

Stokes (2014) described ten principles for developing a methodology (Box 3.1). With this in mind, assessments should include:

- Adequate and verifiable data and information: biomass should be considered a commodity like other agricultural and forest products. Investments are needed to provide such information.
- Yield: a significant variable in biomass supply is yield either from residues and wastes or from energy crops. Geography and local climate alone create variability. The literature, empirical studies and expert opinion are used to develop yield estimates. Scenarios incorporated a range of annual yield increases.
- Supply curves: estimates for biomass availability assume different prices. Farm gate/roadside costs are developed for each feedstock and modelled to determine biomass availability at a given price.
- Sustainability: this is another important, underlying premise to be incorporated into the analysis. Different feedstocks require different approaches. These include using multipliers and coefficients to model certain parameters such as soil carbon retention.
- Land availability and land-use change: land availability is important in estimating biomass production and land-use change is an important sustainability issue. Land competition between conventional crops and energy crops, and among energy crops are modelled.

Box 3.1. Ten principles for developing a methodology to estimate biomass potential

1. Determine desired outcomes and probable uses; available data and analytical resources; and, then determine the "best" approach.
2. Use commonly accepted terminology and definitions of land-use classes and other variables and functions. Be consistent.
3. Use well- and consistently-defined feedstocks – from categories to a single feedstock.
4. Use various analytical tools dependent on availability of data and models; document and explain.
5. Use various data sources (mostly publicly available for transparency) and document extrapolation; rely on many disciplines and professionals to have the technical depth required to understand and use the data appropriately.
6. Use additional data, analyses and experts so that scenarios are both realistic and useable.
7. Put other models to work to overcome specific issues such as sustainability criteria.
8. Work at the most appropriate spatial level based on data and models. Try to complete analysis for smallest spatial units and aggregate upwards to area, state, region and national.
9. Provide and document all background work and assumptions.
10. Explain and document the details of the analyses and the outcomes and the application of the results.

Source: Stokes (2014), "The 'Billion Ton Update': Methodologies and implications".

A regional example in the United States

Dedicated energy crops and crop residues are considered to be able to meet herbaceous demands for the new bioeconomy in the central and eastern United States. Perennial warm-season grasses and corn stover are well-suited to the eastern half of the country. They provide opportunities for expanding agricultural operations in the region. The Department of Agriculture's Agricultural Research Service and collaborators associated with its Regional Biomass Research Centers have developed a suite of warm-season grasses and associated sustainable management practices. Second-generation biofuel feedstocks provide an opportunity to increase production of transportation fuels from recently fixed plant carbon rather than from fossil fuels. Although there is no "one-size-fits-all" bioenergy feedstock, crop residues like corn stover are the most readily available bioenergy feedstocks. However, on marginally productive cropland, perennial grasses provide a feedstock supply while enhancing ecosystems services. Twenty-five years of research have demonstrated that perennial grasses like switchgrass are profitable and environmentally sustainable on marginally productive cropland in the western corn belts and southeastern United States (Mitchell et al., 2016).

Harmonising sustainable biomass potential

The *Billion Ton* reports may give leads on how to harmonise the approaches, which, as already highlighted, vary in underlying methodologies, assumptions and analyses. It is important to estimate effectively the sustainable capacities for biomass production for both domestic use and international biomass trade.

Japan and biomass policy

Biomass availability is an issue for the development of a Japanese bioeconomy. However, Japan was one of the earliest developers of major biomass policy, which Table 3.1 charts from 2002. Other OECD countries could learn from Japan, especially considering its success in creating "biomass towns". Japan's practical experience in making value chains may also be transferrable.

Table 3.1. **Japanese biomass policies 2002-12**

Year	Policies	Outline
2002	Biomass Nippon Strategy	– Basic national strategy to realise sustainable society with full biomass utilisation, and beginnings of Biomass Towns 2004
2005	Kyoto Protocol Target Achievement Plan	– Promoting widespread use of biofuels – Building Biomass Towns and developing biomass energy conversion technologies
2006	Biomass Nippon Strategy (revised)	– Biomass energy for fuels for transportation – Goal of 300 Biomass Towns by 2010
2009	Basic Act for the Promotion of Biomass Utilisation	– Planned promotion of biomass utilisation policy – Drawing up National Plan for Promotion of Biomass Utilisation – Setting up National Biomass Council
2010	Basic Energy Plan	– Introduced renewable energy in 10% primary energy supply by 2020
2010	Act Concerning Sophisticated Methods of Energy Supply Structure	– Required oil refiners to produce specified volumes of biofuels
2010	National Plan for Promotion of Biomass Utilisation	– Setting targets for 2020 – Setting basic policies on technology development for biomass utilisation
2012	Biomass Industrialisation Strategy	– Specified targeted conversion technologies and biomass for realising biomass industrialisation – Setting principles and policies for realising biomass industrialisation

Measuring biomass sustainability

No internationally agreed tools or indicators for biomass sustainability

There are no internationally agreed tools or indicators to measure biomass sustainability. Life cycle analysis (LCA) is frequently discussed as a tool, but only considers environmental performance, and not economic or social factors. Moreover, significant data gaps exist in the availability of life cycle inventory data (Grabowski et al., 2015). Other sustainability tools fail to meet fundamental scientific requirements for index formation: normalisation, weighting and aggregation (Böhringer and Jochem, 2007). No one assessment tool fits the needs of biomass sustainability.

There is also no international agreement on criteria to measure biomass sustainability. International harmonisation requires not only robust analysis, but also consensus, which is often more difficult to achieve. Social criteria are sometimes regarded as unreliable and impractical because they are difficult to measure. As a result, they tend to be assigned a low ranking (van Dam and Junginger, 2011). But they may have strong bearing on true sustainability by analysing issues such as workers' rights and land rights (Shawki, 2016).

As their major limitation, the vast majority of methods cannot aggregate the different sustainability issues into a single measure objectively (Gaitán-Cremaschi et al., 2015). Aggregation requires making complicated trade-offs between sustainability and other factors that are not necessarily intuitive. Practitioners can only generate an overall sustainability number by using their own weighting factors when aggregating the different impact categories; this introduces subjectivity.

LCA in assessment of biomass sustainability

LCA methodology has unique advantages when analysing the environmental performance of products. In theory, based on accounting for all relevant material flows throughout the entire life cycle, it allows a complete picture of certain environmental burdens associated with a product. This enables comparisons across technological boundaries and permits identification of relevant stages in the life cycle, as well as improvement options.

However, LCA methodology has fundamental shortcomings, including dependency on numerous subjective choices, need for simplifications, lack of adequate data and limited precision. These limitations cannot be overcome by another layer of rules in addition to existing standards; they are inherent in the system of life cycle assessment. The lack of a standardised accounting for the biogenic carbon storage in bio-based materials presents a key challenge to LCA practitioners (Pawelzik et al., 2013).

In addition, LCA is not the definitive tool to suitably characterise all environmental impacts. Many impacts cannot be reasonably related to reference flows because the effects depend on space, time and threshold. Sound environmental assessments require a mix of different tools (e.g. environmental impact assessment, human health and environmental risk assessment, technology assessment). These tools must take due account of their strengths and weaknesses.

LCA is suitable for orientation of certain aspects at the onset of developing indicators or setting regulatory requirements. It delivers rough estimates rather than precise figures. However, suitable production, consumption or disposal indicators are typically more robust, more meaningful or relevant, and cheaper. They can also be measured and are easier to verify (or to enforce).

Harmonised methodologies to calculate the environmental footprint (EF) of products have been developed. EF methodologies are by no means new; rather, they constitute a remix of existing tools and related guidance. A key concept for improving comparability is the development of "Product Environmental Footprint Category Rules" (PEFCRs) (European Commission, 2016) for specific products. These are being tested over three years in the European Union with the help of volunteer stakeholders and industry (European Commission, 2017). The objectives of the EF pilot phase are the following:

- Set up and validate the process of the development of product group-specific rules (PEFCRs), including the development of performance benchmarks.
- Test different compliance and verification systems to set up and validate proportionate, effective and efficient compliance and verification systems.
- Test different business-to-business and business-to-consumer communication vehicles for Product Environmental Footprint information in collaboration with stakeholders.

A framework for indicator development embedded in the system of political decision making would also be useful. This could translate priority environmental concerns and broad target setting into specific quantified environmental demands. It would do this at the country or region level (e.g. European Union), as well as at organisational and product levels. A useful starting point for a harmonised methodology would include a discussion of the pros and cons of current practices. On this basis, policy makers could identify needs for improvement covering all dimensions of the subject in question.

International harmonisation and a level playing field for biomass sustainability

Biomass sustainability assessment needs to be harmonised internationally. Assessments are a patchwork of voluntary standards and regulations with a lack of comparability. In a survey of 11 European countries (Knudsen et al., 2015), 8 saw the need for a more consistent and standardised approach to sustainability criteria across the different bioeconomy fields. This need covers widely different criteria and indicators, voluntary schemes and EU-level approaches. The general arguments for a uniform approach to sustainability criteria are to increase transparency, avoid market distortions and enable comparisons across countries.

Much of the biomass shipped internationally is for bioenergy. This risks too much attention on only one part of the bioeconomy and only the energy transition, distorting the playing field even further. Different fields of the bioeconomy are expected to interact. For example, the cascading use of biomass (Odegard et al., 2012; Keegan et al., 2013) envisages the same biomass in use for high- and low-value chemicals and materials, biofuels and bioenergy. A common, level playing field for all sustainable biomass uses is needed (Carus et al., 2014). This is vital for the economic operation of integrated biorefineries.

Policy implications

- LCA is an environmental tool that does not address economic and social impacts. However, these impacts are crucial for policy decisions, particularly where such impacts are vital. This seems to indicate the need for a fundamental review of LCA's utility in biomass sustainability assessment.
- Social impacts especially are difficult to quantify and are therefore easily sidelined. The most robust indicators must be carefully identified. Qualitative indicators (e.g. compliance with organic farming standards) merit inclusion in environmental assessment.
- Complementing and/or alternative environmental assessment approaches could be considered. These could involve indicators tailored to specific product groups that are relevant, robust, verifiable and cost-effective.
- An adequate forum with a broad range of stakeholders for the critical review of LCA methodology and possible alternative approaches for product assessment could be identified.

Is the market more able to provide a unified approach to biomass sustainability assessment?

An "index" approach requires expressing multiple input-output variables with a common denominator. Such an approach helps integrate and compare sustainability issues affecting human well-being at different temporal and spatial scales. One common denominator that the market understands is money. This would involve monetising the "good" and "bad" inputs and outputs. Importantly, the analysis would have to incorporate several sustainability issues into a single measure of sustainability.

Gaitán-Cremaschi et al. (2014) suggested the total factor productivity (TFP) approach to the problem. TFP reflects the rate of transformation of inputs (capital, labour, materials, energy and services) into outputs (biomass stock). In this case, negative social and ecological externalities associated with different sustainability issues are included in terms of "bad" outputs.

The TFP index would use prices that reflect the relative importance of input and output variables towards sustainability. In this solution, observed prices can be used for the marketable inputs and outputs. Shadow prices need to be estimated for externalities that are non-tradable in conventional markets. As a result, related price information does not exist. In other words, the TFP index would use (shadow) prices[1] to reveal the relative performance of a biomass production chain reflected in the form of price signals.

Thus, a biomass chain with the best sustainability performance – the highest TFP score – would produce the highest ratio of output to input where the "bads" are output penalties that lower the sustainability performance. Moreover, the TFP index could compare multiple chains with different sets of outputs and inputs.

Purported advantages of the TFP approach

- It includes externalities (social and economic).
- Numerical harmonisation allows aggregation into a common metric.
- Inputs, outputs and bad outputs are converted to a common, universally understood unit: money.
- Access to market price data makes policy negotiations easier – prices are tangible, while qualitative indicators such as child labour are not.

Policy implications

- The acceptance of such a tool would require consulting with all stakeholders (policy makers, business stakeholders, non-governmental organisations [NGOs]) on:
  - the selection of sustainability issues (i.e. the inputs and outputs)
  - the method for aggregating multiple input and output variables in the TFP index.
- The application of the TFP index would require a common base level of understanding of sustainability. This, in turn, would have to be defined from regional, national and/or international biomass sustainability debates. In this way, inputs and outputs could be selected around issues of sustainability that are of established concern for expert scientist communities, policy makers and the well-being of society. These include, for example, global warming, energy, innovation, human rights, equity and land use.
- The aggregation methodology would have to be agreed upon and accepted internationally. Aggregating sustainability issues using price information can benefit policy makers in data-poor situations, where information about different sustainability issues is still lacking. Nevertheless, it requires decisions about the importance of different sustainability issues expressed in the "true" shadow price. These decisions imply incorporating social, political and ethical values in monetary terms. These values often conflict, and could be deeply contentious in society. This would require careful handling and transparent stakeholder communication. Economic evaluation tools can help estimate shadow prices for decision making.
- The other approach to aggregation, using distance functions, allows easily integrating multiple environmental and social externalities without requiring (shadow) prices. Nevertheless, it must include a large set of observations for the multiple inputs and outputs in the sustainability assessment.

ILUC: Where food and non-food uses of biomass collide

There is a direct land-use change (LUC) where previously uncultivated land is used to grow crops for industrial use. In this case, there are protocols to calculate the GHG impact of LUC. The protocols are used, for example, in the Renewable Energy Directive (RED). Perhaps the most controversial issue regarding bio-based production from biomass is indirect land-use change (ILUC); this occurs when land for food production is converted to grow a crop for non-food use. It is assumed that food production is essential and that the lost food production will be diverted elsewhere. Using previously uncultivated land causes large initial increases in GHG emissions e.g. by encouraging deforestation. Since a primary purpose of biomass for industrial use is to reduce GHG emissions, the impacts of ILUC should be considered.

As an example, the UK Government's Gallagher Review (Renewable Fuels Agency, 2008) stated that biofuel policy must address ILUC to have clear climate benefits. However, its measurement is extremely complex, and some would contend impossible. Further, uncertain conditions undermine investor confidence, which affects the political viability of biofuels.

Political progress on ILUC has been slow with Europe – a good example of divided opinion. ILUC was considered to be inadequately addressed in both RED and the Fuels Quality Directive (FQD). As a result, some biofuels may consequently have few environmental benefits compared with fossil fuels. Indeed, they may even increase GHG emissions rather than generate net savings. In 2012, to address ILUC, the European Commission proposed a directive amending the RED and FQD. It was subsequently adopted by the Council and Parliament, and published in September 2015 (Europa, 2015). In it, fuel suppliers and the European Commission are to report on emissions deriving from ILUC. However, these emissions are not included in the sustainability criteria for the biofuels or the GHG calculation methodology of the RED and FQD. Implementation of the ILUC Directive has been slow. This is partly because it is still quite new, but also because EU member states hold different positions (CE Delft, 2015).

What can be done to ease tension between food and non-food uses of biomass?

Promoting uses of biomass that are unlikely to have a large impact on ILUC is one alternative to the tension between food and non-food uses of biomass. This would provide a means of mitigating ILUC, while avoiding the need for relying on controversial modelling results. In essence, to demonstrate a low ILUC impact, biomass needs to prove the feedstock has not come from land in competition with food production or from carbon-rich lands (forests, peat lands).

Mitigation options that use supply chain certification schemes could provide a workable solution for addressing ILUC. Such a process could allow developers to provide evidence that their biomass for industrial uses has minimal ILUC impact. For example, they could use abandoned or degraded land, or improve crop yields. As such, they should be exempt from application of any ILUC penalty, such as an ILUC factor. Policy makers could build upon this concept to provide a more satisfactory outcome to addressing ILUC in policy.

Policy implications

- ILUC modelling is in no state to be used in policy making relating to biomass sustainability.
- All forms of biomass could be acceptable as feedstock for the bioeconomy; this could be mirrored in public debate and perception, as well as in specific policies.

- Biomass must meet established international sustainability standards covering GHG savings, sustainable land use and environmental protection. These criteria could be integrated into supply chain certification schemes.
- Public financial incentives should only be based on higher resource and land-use efficiencies, sustainability and GHG savings and the lowest possible level of competition with food.
- Food or non-food biomass should not be taken as the sole acceptance criterion.
- Policies for producing sugar for industry use should be examined. For example, sugar beet is an attractive feedstock for the European chemical industry. It has low impact on the food and feed sector as increased yield is decreasing areas under cultivation.
- Added value, employment and innovation speak in favour of supporting industrial use of biomass for materials and chemicals. This would replace disproportionately allocating biomass to fuels and energy applications. Greater value added can only improve on ILUC calculations and implications relating to biomass sustainability.

Does the use of marginal land alleviate the complexities of sustainability?

Sustainable biomass and marginal land

Large quantities of food and/or feed crops such as corn and soybean are used to produce grain-based ethanol and biodiesel. While cultivating highly productive crops on prime agricultural land can produce large quantities of biofuels, it can also harm the environment. Along with other factors, the practice could contribute to rising food prices as well (i.e. the food vs. fuel debate).

An alternative approach is to grow lignocellulosic (or cellulosic) crops on "marginal lands". Marginal land may be defined as follows: land not used for food production because of some inherent limitation; low fertility, highly erodible, or otherwise not suitable for annual crops and not used for grazing. Growing cellulosic feedstocks on such lands is advantageous due to the low management intensity required, increased soil carbon stocks, and reduced soil erosion and GHG emissions.

There are two main challenges to achieving this:

1. Choosing the right crops to ensure sufficient productivity with environmental benefits: achieving sufficient yields on inherently unproductive lands requires choosing plants that can grow well on marginal soils.
2. Understanding the landscape dynamics that influence the supply and distribution of feedstocks: growing biofuel feedstocks on marginal lands may further amplify the complexity of feedstock supplies. Parcels of marginal lands might be spread across landscapes. They may or may not be connected by a suitable road network, or be large enough for successful harvesting and handling of biomass. Transport, management and biodiversity implications need to be understood.

Gelfand et al. (2013) identified 35 locations across the north-central United States where biorefineries with production potential above 133 million litres could be built. These biorefineries could produce ~ 21 billion litres of cellulosic ethanol per year. By 2022, this will equal about 25% of the mandate for the US Energy Independence and Security Act of 2007.

However, before establishing a sustainable biofuel economy, three questions must be answered.

1. What are the direct and indirect effects of land conversion on GHG emissions? As previously noted, ILUC issues are complex. The models are not ready for use in policy or legislation.
2. What is the availability of marginal lands for biofuel crop production? What is the potential productivity of available lands, and where are they located relative to potential biorefineries? How will this interact or interfere with social issues, such as tourism? In addition, are landowners willing to grow biofuel crops in the first place?
3. What is the ideal biofuel feedstock? For example, what are the trade-offs associated with annual and perennial biofuel crops? Perennial feedstocks provide various ecosystem services such as soil carbon sequestration and stabilisation in addition to the biomass produced. They require a low input of agrochemicals. Further, they have a high ratio of energy return on investment and generate high climate mitigation benefits. And they have potential to produce greater yields than annual plants on marginal lands. However, if the demand changes, other crops could replace annual plants. Perennial crops need to be grown for several years before harvesting is possible; they cannot be rotated as often as annual feedstocks.

An inter-disciplinary approach could support better understanding of public and landowner perspectives. Specifically, it could shed light on use of existing landscapes for renewable energy production as part of more general ecosystem services such as clean water and biodiversity.

Policy implications

- Yield alone does not justify supporting an energy crop. Policy makers should assess additional benefits through, for example, enhanced ecosystem benefits that foster biodiversity.
- Best management practices are needed for biofuel feedstock production. Combining the right crop with the right location and the right cultivation practices can generate maximum environmental benefits. Guidelines for sustainable feedstock production need to be developed and will require monitoring tools for assessment.
- The time dimension should be integrated into assessment of the environmental impacts of biofuel feedstocks. Forest will require decades to grow back and to uptake CO_2, which will be released due to harvest and use of forest biomass as a biofuel feedstock. Harvesting of existing mature forests therefore is not providing expected climate mitigation.
- Best management practices can help select suitable marginal lands and implement the growth of cellulosic feedstocks on them. Although they are potentially less productive than high-input/high-yield crops, such feedstocks can provide more environmental benefits, which would need to be monitored.
- Development of breeding and selection programmes for new feedstock crops should be supported.
- Implementation of low-input cropping systems, such as grasses, should have high priority.

- A spatial inventory of lands in areas suitable for biofuel production is needed to inform development of land-use guidelines.
  - Include land connectivity and assessments of potential yields. This must identify existing land-use patterns at a small spatial scale to be relevant for the growth of feedstocks, as an alternative land use (i.e. sub-kilometre).
  - Impacts of agricultural intensification are experienced domestically (i.e. direct land-use change) and globally (i.e. indirect land-use change), and both should be considered.

Technology tools

Lynch et al. (2013) suggest that forests are best monitored through satellite technology. An interesting development is the combination of machine vision software and light detection and ranging (liDAR) technology by Arbonaut of Finland. Flying at an altitude of around 2 kilometres, laser beams can generate three-dimensional point cloud data on an object as small as a single tree on the ground. And knowing the diameter of the crown of the tree, its volume can be predicted (Ministry of Economic Affairs and Employment of Finland, 2017). Making such forestry inventories supports sustainable forestry management. The technology can also be used to assess carbon stocks in tropical forests. It can calculate the amount of CO_2 removed from the atmosphere, entitling a country to payments for carbon capture via forests under the Paris Agreement.

Note

1. Shadow price is the opportunity cost of an activity or project to a society, computed where the actual price is not known or, if known, does not reflect the real sacrifice.

References

Böhringer, C. and P.E.P. Jochem (2007), "Measuring the immeasurable – a survey of sustainability indices", *Ecological Economics*, Vol. 63, Elsevier, Amsterdam, pp. 1-8.

Bosch, R. et al. (29 July 2015), "Define biomass sustainability", Nature blog, Vol. 523, Nature Publishing Group, London, pp. 526-527.

Carus, M. et al. (2014), "Proposals for a reform of the Renewable Energy Directive to a Renewable Energy and Materials Directive (REMD). Going to the next level: Integration of bio-based chemicals and materials in the incentive scheme", *Nova Paper on bio-based economy*, No. 7, Nova Institute, Hürth, Germany.

CE Delft (2015), "Assessing progress towards implementation of the ILUC Directive", CE Delft, The Netherlands, www.cedelft.eu/en/publications/download/2014.

Dornburg, V. et al. (2008), "Assessment of global biomass potentials and their links to food, water, biodiversity, energy demand and economy", *Report*, No. 500102 012, Scientific Assessment and Policy Analysis for Climate Change (WAB), Bilthoven, Netherlands.

Europa (2015), Directive 2015-2013, Eur-Lex, http://eur-lex.europa.eu/legal-content/EN/TXT/?uri=celex%3A32015L1513 (accessed 7 December 2017).

European Commission (2017), "The Environmental Footprint Pilots", webpage, http://ec.europa.eu/environment/eussd/smgp/ef_pilots.htm (accessed 7 December 2017).

European Commission (2016), "Guidance for the implementation of the EU Product Environmental Footprint (PEF) during the Environmental Footprint (EF) pilot phase", Version 5.2, February 2016, European Commission, Brussels.

Gaitán-Cremaschi, D. et al. (2015), "Benchmarking the sustainability performance of the Brazilian non-GM and GM soybean meal chains: An indicator-based approach", *Food Policy*, Vol. 55, Elsevier, Amsterdam, pp. 22-32.

Gaitán-Cremaschi, D. et al. (2014), "Measuring biomass production performance towards sustainability", *Working Paper*, Business Economics Group, Wageningen University, The Netherlands.

Gelfand, I. et al. (24 January 2013), "Sustainable bioenergy production from marginal lands in the US Midwest", Nature Letter, Vol. 493, Nature Publishing Group, London, pp. 514-520.

Grabowski, A. et al. (2015), "Life cycle inventory data quality issues for bioplastics feedstocks", *International Journal of Life Cycle Assessment*, Vol. 20/5, Springer, pp. 584-596.

Keegan, D. et al. (2013), "Cascading use: A systematic approach to biomass beyond the energy sector", *Biofuels, Bioproducts and Biorefining*, Vol. 7, Wiley Online Library, pp. 193-206.

Knudsen, M.T. et al. (2015), "Mapping sustainability criteria for the bioeconomy", *Report*, Department of Agroecology, Aarhus University, Denmark.

Ladanai S. and J. Vinterbäck (2009), "Global potential of sustainable biomass for energy", *Report*, No. 2009:013, Department of Energy and Technology, Swedish University of Agricultural Science, Uppsala, Sweden.

Lynch, J. et al. (2013), "Choose satellites to monitor deforestation", *Nature*, Vol. 496, Nature Publishing Group, London, pp. 293-294.

Ministry of Economic Affairs and Employment of Finland (2017), "Wood-based bioeconomy solving global challenges", Helsinki, Finland.

Mitchell, R.B. et al. (2016), "Dedicated energy crops and crop residues for bioenergy feedstocks in the central and eastern USA", *BioEnergy Research*, Vol. 9/2, Springer, pp. 384-398.

Odegard, I. et al. (2012), "13 solutions for a sustainable bio-based economy. Making better choices for use of biomass residues, by-products and wastes", CE Delft, The Netherlands, http://docplayer.net/49408699-13-solutions-for-a-sustainable-bio-based-economy.html.

Pawelzik, P. et al. (2013), "Critical aspects in the life cycle assessment (LCA) of bio-based materials – Reviewing methodologies and deriving recommendations", *Resources, Conservation and Recycling*, Vol. 73, Elsevier, Amsterdam, pp. 211-228.

PBL (2012), "Sustainability of biomass in a bio-based economy: A quick-scan analysis of the biomass demand of a bio-based economy in 2030 compared to the sustainable supply", *PBL Note*, No. 500143001, Netherlands Environmental Assessment Agency, Amsterdam.

Renewable Fuels Agency (2008), *The Gallagher Review of the Indirect Effects of Biofuels Production*, Renewable Fuels Agency, St Leonards-on-Sea, East Sussex, United Kingdom.

Rettenmaier, N. et al. (2008), *Status of Biomass Resource Assessments, Version 3*, Biomass Energy Europe, European Commission, Brussels.

Saygin, D. et al. (2014), "Assessment of the technical and economic potentials of biomass use for the production of steam, chemicals and polymers", *Renewable and Sustainable Energy Reviews*, Vol. 40, Elsevier, Amsterdam, pp. 1153-1167.

Schueler, V. et al. (2016), "Productivity ranges of sustainable biomass potentials from non-agricultural land", *Environmental Research Letters*, 11 (2016) 074026, http://dx.doi.org/10.1088/1748-9326/11/7/074026.

Seidenberger, T. et al. (2008), *Global Biomass Potentials – Investigation and Assessment of Data, Remote Sensing in Biomass Potential Research, and Country-specific Energy Crop Potentials*, German Biomass Research Centre, Leipzig.

Shawki, N. (2016), "Norms and normative change in world politics: An analysis of land rights and the Sustainable Development Goals", *Global Change, Peace & Security*, Vol. 28, Taylor & Francis Online, pp. 249-269.

Stokes, B.J. (2014), "The 'billion ton update': Methodologies and implications", presentation at the OECD workshop "Sustainable biomass drives the next bioeconomy: A new industrial revolution?", Paris, 10-11 June.

US DOE (2017), *2016 Billion-Ton Report, Volume 2: Environmental Sustainability Effects of Select Scenarios from Volume 1*, Efroymson, R.M. et al. (eds.), ORNL/TM-2016/727, Oak Ridge National Laboratory, United States Department of Energy, Oak Ridge, United States.

US DOE (2016), *2016 Billion-ton Report: Advancing Domestic Resources for a Thriving Bioeconomy, Volume 1: Economic Availability of Feedstocks*, Langholtz, M. et al. (eds.), ORNL/TM-2016/160, Oak Ridge National Laboratory, United States Department of Energy, Oak Ridge, United States.

US DOE (2011), "U.S. billion-ton update: Biomass supply for a bioenergy and bioproducts industry", Perlack, R.D. and B.J. Stokes (eds.), ORNL/TM-2011/224, Oak Ridge National Laboratory, United States Department of Energy, Oak Ridge, United States.

US DOE (2005), "Biomass as feedstock for a bioenergy and bioproducts industry: The technical feasibility of a billion-ton annual supply", ORNL/TM- 2005/66, Oak Ridge National Laboratory, United States Department of Energy, Oak Ridge, United States.

Van Dam, J. and M. Junginger (2011), "Striving to further harmonization of sustainability criteria for bioenergy in Europe: Recommendations from a stakeholder questionnaire", *Energy Policy*, Vol. 39/7, Elsevier, Amsterdam, pp. 4051-4066.

Chapter 4.

Biotechnology and biomass sustainability

This chapter focuses on biotechnology in food production and the future roles of marine biotechnology. Given the extensive discourse on competing uses of biomass in food and industry, this is an important area for policy makers. If land extensification possibilities are limited, and agricultural productivity is declining, the industrial use of biomass would also be limited. Even as other forms of biomass are being sought as biorefinery feedstocks, agricultural productivity and sustainability need to be improved. The yield increase of the so-called green revolution in modern agriculture from the 1950s is flattening out. In addition, agricultural practices with higher inputs, such as pesticides and fertilisers to ensure high yields, are not considered environmentally sustainable. Therefore, the contributions of biotechnology to land extensification and intensification will be crucial in future. In addition, the marine environment remains a virtually untapped resource.

Introduction

Biotechnology, whether through genetic modification or not, is already making a huge impact on the sustainability of food production. This will continue to rise in importance as the human population increases and fossil dependence decreases. Despite apparent progress, the surface has merely been scratched. Biotechnology in agriculture can still make a huge impact.

Marine biotechnology and the sustainable exploitation of the marine environment are in their infancy. Whether exploiting its biomass or genetic potential, the marine environment will play a major role in a sustainable bioeconomy. It will reduce pressure on land and relieve fears about the competing use of terrestrial biomass for food and industry. Nevertheless, the oceans also have to be exploited sustainably.

Although powerful, genomics does not necessarily involve genetic modification (GM) or synthetic biology. Consequently, the negative societal issues that have haunted GM in many applications can be avoided. Rather, -omics technologies can be applied to animal and plant breeding to make selection of traits much more efficient. This is especially important for trees given the long timescales needed for growth and trait expression.

To exploit its full potential, genomics information needs to be linked to phenotypic characteristics. The availability of well-defined linkage maps and the extent of genetic studies conducted on them vary among different crops. This, in turn, influences the feasibility of any activity related to marker-assisted selection (MAS).[1] MAS can reduce the breeding cycle time significantly (e.g. for cassava from five to two years) and is much more accurate (Ly et al., 2013).

Genomics and biotechnology in food production

Increasing incomes in developing economies have contributed to a large increase in meat and milk consumption. From the beginning of the 1970s to the mid-1990s, the increase in meat consumption in developing countries was almost triple the increase in developed countries. Similarly, the increase in milk consumption was more than twice the increase in developed countries (Delgado, 2003). Naturally, this creates strains on a bioeconomy as less biomass can be devoted to industrial uses. Clearly, new ways are needed to increase food production efficiency in these areas.

Chinese Taipei offers a good example of the shift in diet that occurs with development. In the 30 years from 1959-89, per capita consumption of rice halved, while fish consumption doubled, meat consumption (chicken, beef and pork) quadrupled and fruit consumption quintupled (Huang and Bouis, 1996). Similar patterns were seen in Japan and Korea as household incomes increased.

From a bioeconomy point of view, this trend is negative. Large amounts of crops are produced to feed animals; ruminant production has notoriously high greenhouse gas (GHG) emissions (Table 4.1) and water consumption.

Beef production

The Australian beef industry today sees "unprecedented demand from the entire Asia Pacific, as well as the Middle East" (Kondo, 2014). Previously, demand was mostly from Japan, and then later the People's Republic of China (hereafter "China"). But beef production is resource-intensive in terms of land, feed and water, and also creates large GHG emissions. Policy makers are seeking measures to improve efficiency of beef production, and genomics offers some solutions.

Table 4.1. **GHG emissions associated with various meat production systems**

Product	CO_2 (eq./kg)
Beef	44.80
Idaho and Nebraska beef (average)	33.50
Idaho lamb	44.96
Swedish pork	3.3-4.4
Michigan pork	10.16
Chicken	2.0-4.6
Poultry (US)	1.4
Cod	3.2
Farmed salmon (sea-based, UK)	3.6
Farmed salmon (sea-based, Canada)	4.2
Farmed salmon (sea-based, Norway)	3.0
Farmed trout	4.5
Capture fish (global average)	1.7

Source: Jiménez-Sánchez, G. and J.C. Philp (2016), "The bioeconomy, genomics and society".

Genomics has been propagated as a "paradigm shifting" innovation in livestock production over the last decade. The possibility of predicting breeding values using genomic information has exerted major changes within the dairy cattle industry. This technology is now being used in beef cattle, but the diversity of breeds presents a challenge to develop genomic tools.

There is large scope for the development of novel applications in the livestock sector. These include selection tools for new traits (meat quality, disease resistance, feed efficiency, heat tolerance), animal traceability and parentage verification (e.g. McClure et al., 2013). Scientists are sequencing important animals in the global beef industry to identify variants and to associate those variants with the genetic variation observed across beef populations.

Selecting beef cattle for protein production requires appropriate emphasis on economically relevant traits (ERTs). Most ERTs are quantitative, such as early life growth and carcass quality attributes. These traits are output ERTs that impact revenue. Many ERTs are left out of breeding objectives for several reasons. In some cases, the capacity to collect data in the field does not exist. In other cases, the cost of collecting enough data for a national evaluation programme is too high. Many of these ERTs affect input costs of production such as animal health, feed efficiency and adaptability. These traits are fertile ground for the application of genomics technology. There is also great scope for increased international collaboration in all livestock species (Pollak et al., 2012).

In the near future, artificial reproductive technologies (ART), such as artificial insemination, embryo transfer and in-vitro fertilisation, combined with genomic evaluation (GE) approaches, may fine-tune cattle breeding. On the one hand, GE-improved methods will identify the exact gene alleles desired for a given type of animal. On the other, ART will enable checking for the presence of these favourable alleles in early stage in-vitro produced embryos, making the whole selection and breeding process much more accurate.

Development of specific "genomic-audited" lineages will also likely carry specific and interconnected alleles selected inside the traditional breeds. This will offer better chances to the livestock industry to produce the animal required for each type of application, fostering quantity and quality parameters.

The Department of Agriculture, Food and the Marine of the Irish government has a Beef Data and Genomics Programme (BDGP) for 2015-20 (BDGP, n.d.). The EUR 300 million programme is addressing the widely acknowledged weaknesses in the maternal genetics of the Irish suckler herd. It will make a positive contribution to farmer profitability and reduce the greenhouse gas intensity of Ireland's beef production. As one objective, the programme will place Ireland at the forefront of climate-friendly agriculture. It will also further the positive environmental image of Irish beef production, which is considered to add value to Irish beef in high-value markets around the world.

Milk production

Genomics studies of milk have varied goals, underlining its significance as a human food. Topics include the capacity of milk to manipulate the gut microbiota; manipulation of bovine milk fat; genetic selection for economically important traits; and diagnostics. Genomics has revolutionised the dairy industry (see Hayes et al., 2009). Genetic tests are used to select every bull that sires milk-producing cows. Traditionally, after breeding the bull with a cow, the breeder would wait nine months for calves to be born. It would then take another three years until the calves begin lactating to know whether the bull produces higher milk-yielding offspring. Genetic testing doubles the speed of achieving those same milk production gains (Darcé, 2010).

Molecular mechanisms that create variations in milk components are important processes for human nutrition. As such, they require further investigation. Protein content, for example, is an important economic trait. In dairy cattle, the heritability of milk protein yield has been estimated at 23%. Identifying the polymorphisms contributing to milk traits could enable breeding programmes to increase milk protein yields, with obvious economic and societal benefits. Furthermore, identification of the gene pathways involved would contribute to understanding of the mechanisms that regulate lactation. This could also lead to new approaches to improve milk production.

Raven et al. (2013) produced evidence supporting a role for the RNASE5 pathway in milk production, specifically milk protein percent (and not other traits such as fat content or fertility). The evidence indicated that polymorphisms in or near these genes explain a proportion of the variation for this trait. Moreover, the gene set method applied to the RNASE5 pathway could be used to rapidly assess the role of other emerging pathways and functions with a genetic validation relevant *in vivo*.

Bacterial genomics can improve control of the microbiological safety and quality of food products, which is particularly relevant for milk. The process can keep milk free from pathogenic bacteria and ensure the concentration of spoilage microorganisms is as low as possible (Marco and Wells-Bennik, 2008). Human pathogens detected in raw milk include *Campylobacter jejuni*, enterohaemorrhagic *Escherichia coli*, *Salmonella* spp., *Listeria monocytogenes*, *Bacillus cereus* and *Yersinia enterocolitica*.

Non-pathogenic bacterial species determine milk quality and limit shelf-life by producing off-flavours, unwanted acidification and structure defects. Culture collection and model strains are instrumental for gaining knowledge on the behaviour of food-associated bacteria. However, the behaviour of "wild" strains in a dairy environment can differ significantly from the laboratory-adapted reference strains. Obtaining genome sequences of additional dairy isolates can help better understand their survival, persistence and pathogenic potential.

The International Milk Genomics Consortium aims to accelerate understanding of the biological processes underlying mammalian milk genomics (http://milkgenomics.org). It organises the annual International Milk Genomics and Human Health Symposium to promote advancement of milk genomics, proteomics, metabolomics and bioinformatics knowledge tools. The symposium facilitates communication between scientists, sponsors and others to accelerate progress and identify commercial opportunities.

Chicken as a food source in a bioeconomy

Chickens are a major source of protein in the world, with around 20 billion birds alive today. It was the first livestock species to be sequenced and so leads the way for others (Burt, 2005). As its production is relatively low in GHG emissions, chicken is an excellent food source in bioeconomy terms (Table 4.1). It is also cheaper to produce and less energy-intensive than rearing lamb, beef or pigs.

In parallel with the chicken genome sequencing project (Hillier et al., 2004), a consortium set about identifying single nucleotide polymorphisms (SNPs[2]). The availability of a standard set of 10 000 or more SNPs holds much promise towards identification of genes controlling quantitative trait loci (QTL), including those of economic interest.

During the past 80 years, modern selective breeding has made spectacular progress in both egg and meat production traits. These successes, however, have also generated undesirable traits that impose added costs on the industry. With increased requirement for food safety, there will be a need to reduce the use of chemicals and antibiotics and increase genetic resistance to pathogens. Further, the consumer wants high-quality products, such as increased egg shell strength; such new traits are difficult and costly to measure by conventional genetic selection. Developments in poultry genomics in the last few years promise new solutions to these problems. Therefore, genomics research may be expected to be directed at these "sustainability" criteria.

Figure 4.1. **Selective breeding of poultry for higher meat production and more efficient feed conversion**

(a) Feed conversion ratio over 30 years in meat-producing broilers at 42 days

(b) Broiler feed to produce a 2.5 kg chicken at 42 days

Source: Jiménez-Sanchez and Philp (2015), "Genomics and the bioeconomy: Opportunities to meet global challenges".

One key trait improved every year through selective breeding is feed efficiency (Figure 4.1) – the number of kilos of animal feed needed to produce a kilo of poultry meat (Technology Strategy Board, 2010). Genomic technologies are expected to enhance this trend. Since animal breeding is cumulative, even small enhancements to the rate of improvement can multiply into huge differences for commercial customers over time and have very large impacts. As a result, the same land resources can feed more people. Improved feed efficiency can also free up land as a resource such as to produce biomass for industrial use.

The Aviagen genomics project is concerned with identifying naturally occurring markers within the genome of elite birds (Aviagen, n.d.). It uses those markers to help breed stronger and more productive birds through the selective breeding programme, a completely natural process. Aviagen became the first company to include genomic information as a critical additional source of information in a research and development (R&D) breeding programme.

Fish and food security

Between 1998 and 2008, global exports of fish products doubled to a value of over USD 100 billion. It is estimated that more than 20 000 species of fish are used for food. Of more than 34 million global fishers (i.e. excluding aquaculture) in 2008, over 8.25 million were in China alone. Further, more than 2.25 million were in Indonesia compared to under 13 000 in Norway. Per capita consumption of fish continues to rise – from 10 kg to 19 kg between the 1960s and 2012 (FAO, 2014).

From the bioeconomy perspective, fish protein relieves pressure on land as the source of biomass for both agricultural and industrial uses. Given the health benefits and the lower GHG emissions associated with fish (Table 4.1), eating more fish would appear to be desirable for a future bioeconomy.

Overfishing has reduced some fish stocks to near extinction. Further, destructive fishery practices, such as bottom trawling, have damaged the habitat of the ocean floor. Coastal development and the resulting domestic and industrial wastes continue to perturb marine ecosystems and threaten coastal habitats in some areas. In extreme cases, agricultural pollution has resulted in hypoxia, which weakens established ocean ecosystems and sometimes leads to permanent "dead zones".

About 90% of global wild fish stocks are already at capacity or are in precipitous decline (Tinline, 2015), with 60% of wild stocks fully fished and 30% overfished (FAO, 2014). Wild fisheries should therefore be regarded as "not necessarily renewable". Well-reported universal difficulties associated with wild fisheries are related to identifying fish species. These include species with limited diagnostic morphological features, cryptic species, juvenile identification or unavailability of adequate drawings and descriptions.

Aquaculture has continued to grow in volumes and species as a consequence of deeply troubled wild fishing and increasing demand for fish. It produced 66.6 million tonnes in 2012 compared to 91.3 million tonnes for wild fish. However, the growth rate of aquaculture is more instructive. From 2007-12, the aquaculture industry grew by 33.5%, whereas capture fishing grew by a mere 0.6% (which is effectively static). Continuing future increases in demand are likely as the global middle class explodes in growth (D'Hondt et al., 2015). Most future growth in fish will likely have to come from aquaculture.

The benefits arising from the rapid growth in aquaculture have been accompanied by serious environmental, social and production challenges. For most countries, reliance on fish feeds remains an issue as they are often derived from scarce wild resources. Fish health, rearing and containment are also constant challenges. To grow and fulfil the promise of a "blue revolution", aquaculture will need to balance its long-term environmental sustainability with its goal of growing large fish rapidly. Marine biotechnology offers some solutions.

Biotechnology in the capture fish and aquaculture industries

The increased availability of hardware, software and genetic information in the last decade has opened possibilities for biotechnology to contribute to both capture fishing and aquaculture. Many of these applications are not related to GM technologies and therefore are unlikely to cause public resistance.

Capture fishing: Traceability is becoming urgent

Almost 34% of the world's fisheries catch from 1950-2002 lacked species-level identification. The use of DNA barcodes for species delimitation, and the availability of a standardised and globally accessible database (BOLD, n.d.), facilitate numerous related applications. These include issues relating to traceability, eco-labelling, illegal fishing and fish fraud (Costa et al., 2012), and more fundamental information such as migration and dispersal behaviour (Box 4.1). A common fraudulent practice is species substitution. This can be unintentional or intentional to evade taxes, launder illegally caught fish or to sell one fish species for a higher-priced species. Illegal, unreported and unregulated (IUU) fishing remains a major threat to marine ecosystems (FAO, 2014). Traceability is becoming urgent (Waughray, 2017).

Box 4.1. Atlantic herring identification through genomics

Herring has been an important food source for hundreds of years over a very wide distribution. For many decades, it has been in dramatic decline. The Pacific herring, however, has sustained low abundances even after reductions of fishing pressure. Offered reasons include climate-induced ecological changes in distribution of predators and prey; disease; overfishing; and the rebound of marine mammal populations that prey on herring (McKechnie et al., 2014). A study of Atlantic herring on the Gulf of Maine–Georges Bank indicated that predation mortality rates were relatively low during the 1960s, when Atlantic herring were abundant. However, they increased in the late 1970s and early 1980s, when Atlantic herring declined. Predation mortality rates declined in the 1990s as Atlantic herring abundance increased. Sustainable fisheries management for herring is therefore extremely complicated. In addition, herring are generally highly migratory (Overholtz et al., 2008). This further complicates fish stock management, which requires precise and accurate data on the population identity of harvested fish. This allows managers to maximise long-term fisheries yield at minimal risk to population viability.

The risk associated with failure to identify individual populations in herring stock assessments is well known. Weak levels of differentiation among populations have prevented accurate assignment of individual fish to specific origins. Genomic resources, especially single nucleotide polymorphisms (SNPs), heralded the identification of Atlantic herring to unprecedented levels of geo-localisation in this weakly structured fish.

Bekkevold et al. (2015) extended the utility of genetic techniques for herring, demonstrating the applicability of Genetic Stock Identification (GSI). They showed this approach can be used as an adaptive tool to address biodiversity indicators applicable to natural resource management, and also to address illegal fishing and mis-labelling.

The practical advantages of DNA-based identification include the ability to use a range of fresh, preserved or highly processed material. This allows non-experts to collect samples with relatively little cost and effort. The method is also transferrable across laboratories and SNP genotyping platforms. It can be readily extended for additional populations and (or) genetic markers.

About 70% of the global tuna fish catch is taken from the Pacific. Most of the 23 tuna stocks are either over-exploited or depleted. Bluefin tuna are unrivalled in popularity, especially in sushi, and the economic value per fish is unmatched by any other species. However, its over-exploitation seriously threatens its future. Some advocate that consumers should avoid eating bluefin tuna altogether. Moreover, prices of yellowfin tuna and Pacific bluefin tuna are drastically different. But if they are used in cooking, it is difficult even for experts to distinguish between them. DNA barcoding therefore holds out promise for various policy goals: to reduce fraud, to play a role in cultivating conscientious consumerism (by helping conserve threatened species) and to regulate by eliminating market ambiguity effectively (Lowenstein et al., 2009).

To date, no one technique can identify species at the molecular level perfectly. However, DNA barcoding analysis is a significant advancement upon previous DNA techniques because it is based on a universal methodology (Hanner et al., 2011). Linking DNA barcoding to a universally accessible, expert-authenticated database of species identification data would arguably address many problems that plague the system of species authentication (Clark, 2015).

Aquaculture and genomics

Aquaculture production has continued to grow annually at around 6-8%. Today, farmed seafood production (around 60 million tonnes) exceeds that of wild fisheries and has significant potential for future growth. World aquaculture is heavily dominated by the Asia-Pacific region, which accounts for roughly 90% of production, mainly due to China. In 2008, 85.5% of fishers and fish farmers were in Asia, compared to 1.4% in Europe and 0.7% in North America (FAO and WHO, 2010). However, much work remains to improve productivity in Asia: fish farmers' average annual production in Norway is 172 tonnes per person; in China, it is 6 tonnes and in India only 2 tonnes.

High priority traits for farmed fish are the development of single sex populations and improving disease resistance. Production of mono-sex female stocks is desirable in most commercial production since females grow faster and mature later than males. Understanding the sex determination mechanism and developing sex-associated markers will shorten the time for the development of mono-sex female production, thus decreasing the costs of farming.

Tilapia

Nile Tilapia is one of the most important farmed species with a production exceeding 2.8 million metric tonnes in 2010. Tilapia farming is increasingly important in Asia, with (at least) Bangladesh, China, Indonesia, Malaysia, Myanmar, the Philippines, Thailand and Viet Nam all producing significant tonnages. Most Asian countries do not export significant amounts of Tilapia, demonstrating its role in food security.

Tilapia is unusual in that intensive commercial production generally requires all-male stocks. This is because males grow faster, but also to avoid uncontrolled reproduction before harvest. A restriction-associated DNA (RAD) sequencing study by Palaiokostas et al. (2013) identified a reduced candidate region for the sex-determining gene(s) and a set of tightly sex-linked SNP markers. Although they could not identify the causative gene(s), no female was mis-assigned using their sex-associated SNPs. This means those SNPs could be of high practical value towards the production of all-male stocks for the Tilapia aquaculture industry.

Salmon genomics

Salmonids, in particular Atlantic salmon, are among the most important aquaculture species. In 2010, farms worldwide produced approximately 1.5 million tonnes of Atlantic salmon, corresponding to a value of just over USD 7.8 billion (FAO, 2010). It is an important bioeconomy species due to the low GHG emissions associated with farming salmon. Outstanding challenges to the industry include reducing unsustainable fish losses, and controlling or eradicating sea lice and salmon Rickettsia septicaemia. These cost the industry hundreds of millions of dollars in lost production.

Genomic resources for Atlantic salmon are among the most extensive of all aquaculture species. They include several genetic maps, a physical map, an extensive Expressed Sequence Tags database of approximately 500 000 tags and several microarrays (Gonen et al., 2014). In June 2014, the International Cooperation to Sequence the Atlantic Salmon Genome (ICSASG) announced completion of a fully mapped and openly accessible salmon genome, which is housed at its own website (ICISB, n.d.). Some expected outcomes of this research are understanding the attacks by viruses and pathogens on salmon and producing new vaccines to reduce losses through disease; applications for food security and traceability and brood-stock selection for commercially important traits; and better understanding of interactions of farmed salmon with wild counterparts.

Selective, marker-assisted breeding of salmon, made possible due to access to the genome, will be more targeted and efficient. This could, for example, select for individuals that are more resistant to disease and parasites. It could also select fish that grow more quickly while being adapted to new feed types. In the longer term, genomic knowledge should help streamline the aquaculture industry, while providing consumers with healthier farmed salmon that are produced with as little environmental impact as possible.

The power of marker-assisted breeding is illustrated by the success in breeding salmon resistant to sea lice infection (University of Glasgow, 2015) or Infectious Pancreatic Necrosis (IPN) virus (World Fishing & Aquaculture, 2014). Not only can the use of pesticides and antibiotics be omitted, but losses are also greatly decreased. Survival rates of salmon in aquaculture of just a few per cent higher translate into major earnings for the Norwegian aquaculture industry, where the annual turnover is NOK 45 billion (approximately EUR 5.6 billion, or USD 7.6 billion) (Science Daily, 2014).

In the case of salmon, then, the power of relatively small public research funding of genomics to transform an industry is clear. In this case, a single tool can address many industry problems, which makes genomics unique as a solution provider.

Vaccines and the end of the antibiotic era

Marine biotechnology, in the form of new vaccines and molecular-based diagnostics, has already helped increase production, reduce the use of antibiotics and improve fish welfare (Sommerset et al., 2005). In many places, the use of antibiotics has plummeted; Norway produces 99% of farmed salmon without antibiotics. In other countries, however, especially developing countries without access to molecular-based tools and technologies, use of antibiotics remains widespread (Cabello, 2006). Genomic and related technologies have also been used to create new DNA-based vaccines for economically important diseases (e.g. Apex-IHN, Novartis, for the treatment of infectious hematopoietic necrosis in farmed salmon) and highly sensitive specific tools for disease detection (Cunningham, 2002).

Molecular aquaculture

The application of new genomic knowledge and technologies to the practice of aquaculture is termed "molecular aquaculture". This helps distinguish it from the more production-oriented activities in aquaculture such as improved feeding systems, cage design and husbandry. Molecular aquaculture is characterised by the incorporation of new omics knowledge, high-throughput genomics technologies and recombinant DNA technology. These technologies have facilitated selective breeding for economically important traits such as body shape or disease resistance.

Molecular aquaculture holds great potential for increasing sustainable food production. This can help meet anticipated increases in global demand through the culture of species such as salmon, Tilapia, shrimp and oysters. However, molecular aquaculture is developing and diffusing at different rates in different countries, potentially limiting the productivity gains and sustainability of the endeavour.

Hybrid technologies for aquaculture/agriculture

Building solar-powered desalination plants for aquaculture in hot, sunny climates is a possibility (Palenzuela et al., 2015). A hybrid system combining solar-powered desalination with a "floating farm" has been described in concept (Moustafa, 2016). This concept envisages growing crops that need freshwater rather than seawater. Construction of such a system (a "bluehouse" rather than greenhouse) offshore would relieve pressure on land. In other words, it is about moving terrestrial crop production offshore.

Genetic modification in aquaculture

Much more controversial than genomics and traditional vaccine development, GM technology has already been applied to salmon breeding. GM technology varies widely in its acceptability in different countries. In the United States, the Food and Drug Administration (FDA) has approved the first GM salmon. It contains a growth hormone gene from a related species that allows the salmon to grow to market size in 16-18 months rather than over three years. As a result, this salmon consumes at least 25% less feed over its lifetime than conventionally farmed salmon. It thus saves money, which could mean lower prices for the consumer if it reaches the market.

The US FDA began its review of this technology (AquaBounty, n.d.) in 1993. In 2012, it concluded that AquAdvantage was safe for human consumption and unlikely to damage the environment. A lengthy public consultation ensued (Baehr, 2014), during which the salmon was not available to consumers. Finally, on 19 November 2015 it became the first genetically engineered animal to be approved for human consumption in the United States (Ledford, 2015).

Feeding the fish that feed humans

Well-described benefits of eating oily fish come from the long-chain polyunsaturated fatty acids (LC-PUFA) commonly referred to as omega-3 fatty acids. They are, however, not synthesised by the fish themselves; rather, the synthesis is done by single-celled algae and the molecules then pass up the food chain to small, herbivorous fish and thence to large, carnivorous ones (The Economist, 2015).

Paradoxically, farmed fish such as salmon are fed fish meal, made from wild-caught oceanic species such as anchovies that are not in great demand as human food. This helps boost their levels of the two critical omega-3 fatty acids, docosahexaenoic acid (DHA) and eicosapentaenoic acid (EPA). The aquaculture industry needs to find new fish food sources, particularly to replace or supplement these high-quality inputs from fishmeal and oil; lack of new fish food is increasingly seen as a limitation for future growth in aquaculture production (McAndrew and Napier, 2011).

More than one strategy to find new fish food sources is being investigated. First, it is possible to genetically engineer plants to modify seed oil composition[3] to include omega-3 PUFAs; the collective data available (e.g. Usher et al., 2015) confirm the promise of transgenic plants. However, while levels of EPA achieved are equivalent to marine sources, DHA still represents a challenge.

Rather more controversial would be to engineer the fish to make omega-3 fatty acids directly. In one study, zebrafish were transformed with a salmon desaturase, leading to modestly increased tissue levels of EPA and DHA (Alimuddin et al., 2005). While zebrafish is a model organism, Tilapia is an aquaculture species. The Institute of Cellular and Organismic Biology in Chinese Taipei is examining how metabolic engineering in Tilapia could express high levels of omega-3 PUFA biosynthesis genes in the liver.

Crop genomics and precision crops

Feeding the 9 billion people on the planet by 2050 is a major food security issue. Moreover, the demand for biomass for bio-based production of fuels, chemicals and plastics will further stress land availability and productivity. The effects of climate change will exacerbate difficulties facing conventional agriculture. Although seldom acknowledged in discussions of agricultural genetic resources, soils are the critical life-support surface on which all terrestrial biodiversity depends. Meanwhile, the world is losing soil at a rate 13-80 times faster than soil is being formed. In October 2017, the environment secretary of the UK government warned that the United Kingdom is 30-40 years away from "the fundamental eradication of soil fertility" (van der Zee, 2017). In the face of soil destruction, more crops will have to be grown more efficiently. At the same time, methods are needed to halt or limit soil destruction.

Many applications of genomics to crop production will be used in the future bioeconomy e.g. pest resistance, more "efficient" plants that use less water, resistance to environmental stresses and the development of crops that can fix nitrogen to replace synthetic fertilisers. Heat and drought stress are used as examples of applying genomics to agriculture. On the other hand, too much water in the case of rice can also destroy crops.

Heat and drought stress: An increasingly important problem associated with global warming

Improvement of dual stress tolerance to heat and drought in crop plants has become a top priority for the development of agricultural biotechnology for both food and bioenergy markets. Performance Plants, a Canadian company, has identified and completed functional studies of a subset of target genes that constitute a novel regulatory cascade that controls plant responses to the combined stress. In laboratory conditions, *Arabidopsis* and canola plants with missense expression of these regulatory genes could tolerate independent higher temperature or drought treatment. More importantly, when both stresses were applied simultaneously, these plants produced higher seed yield compared to their controls. Charge-scale, multiple season and location field trials further confirmed the dual stress tolerance and yield enhancement properties of the transgenic plants. These results represent a significant breakthrough in crop improvement. Technologies derived from this research could enable farmers around the world to maintain higher yield and productivity over variable and adverse environmental conditions.

Sunflower is mainly grown for edible oil, but is also used for biodiesel, biogas, wood and charcoal, with its use likely to expand. Drought is the main threat to sunflower production, and genomics is taking its place to select new varieties for future crops. It is already rather drought-resistant and the mining of the genome of extremophile variants holds promise for improved drought resistance and increased oil yield (Badouin et al., 2017).

Oil palm genomics

Oil palm illustrates a classic bioeconomy dilemma. It is the most productive oil-bearing crop, accounting for 33% of all vegetable oil and 45% of edible oil worldwide. Although it is planted on only 5% of the total world vegetable oil acreage, increased cultivation competes with dwindling rainforest reserves. Global production of palm oil more than doubled between 2000 and 2012 (FAO, 2013). Thus, the competing imperatives of a bioeconomy are clear to see: creating economic growth while reining in detrimental environmental effects to create a future economy that is sustainable.

Palm oil production is central to the economy of Malaysia, employing close to half a million people. Historical statistics indicate that Malaysian palm oil yields have typically appreciated over time. In 2009, however, an unexpected break in the long-term national growth pattern occurred, which has persisted. Explanations for the abrupt change are varied, which include a combination of adverse weather, ageing trees and plant disease (USDA Foreign Agriculture Service, 2012).

Data indicate the vast majority of trees has already reached or passed through its peak yielding years. A small but growing problem is a lethal fungal disease. *Ganoderma* has the capacity to cause significant yield losses well before it has killed an oil palm. Once it has been introduced, its spores can spread to ever increasing areas of a plantation. Therefore, increasing oil yield and disease resistance would be obvious targets for genomics applications. With growing needs for edible and biofuel uses, increasing yield for oil palm would reduce its rainforest footprint.

The oil palm genome and oil yield

Many of the economic, social and environmental concerns surrounding the bioeconomy are present with oil palm. This puts the crop at the front line of issues around its sustainability. An incident in 2015 exemplifies the seriousness of the concerns. In August 2015, four large groups of Asian companies were excluded from the Norwegian sovereign wealth fund over instances of deforestation in Indonesia (Milne, 2015).

Singh et al. (2013b) published the oil palm genome sequence, which enables the discovery of genes for important traits, as well as alterations that restrict the use of clones in commercial plantings. The oil palm, largely undomesticated, is an ideal candidate for genomic-based tools to harness the potential of this remarkably productive crop. The authors claim that the dense representation of sequenced scaffolds on the genetic map will help identify genes responsible for important yield and quality traits.

The modern oil palm tree *Elaeis guineensis* has three fruit forms: *dura* (thick-shelled); *pisifera* (shell-less); and *tenera* (thin-shelled) (Figure 4.2). The *tenera* palm yields far more oil than *dura*, and is the basis for commercial palm oil production in all Southeast Asia. In 2013, a remarkable discovery was made. The *Shell* gene has proven extremely challenging to identify in oil palm, given the large genome, long generation times and difficulty of phenotyping in experimental populations. Singh et al. (2013a) identified the gene and determined its central role in controlling oil yield. Regulation of the *Shell* gene will enable breeders to boost palm oil yields by nearly one-third. This is excellent news for the industry, the rainforest and its champions worldwide, and also for bioeconomy policy makers.

Figure 4.2. **Oil palm tree fruit forms**

Note: The *Shell* gene is responsible for the oil palm's three known shell forms: *dura* (thick); *pisifera* (shell-less); and *tenera* (thin), a hybrid of *dura* and *pisifera* palms. *Tenera* palms contain one mutant and one normal version, or allele, of *Shell,* an optimum combination that results in 30% more oil per land area than *dura* palms.

Source: Singh et al. (2013a), "Oil palm genome sequence reveals divergence of inter-fertile species in old and new worlds".

Seed producers can now use the genetic marker for the *Shell* gene to distinguish the three fruit forms in the nursery long before they are field-planted. Previously, it could take six years to identify whether an oil palm plantlet was a high-yielding palm. Even with selective breeding, 10-15% of plants are the low-yielding *dura* form due to uncontrollable wind and insect pollination, particularly in plantations without stringent quality control (Tarr, 2013).

Such accurate genotyping has a critical implication for a bioeconomy. Enhanced oil yields will optimise and ultimately reduce the acreage devoted to oil palm plantations. This provides an opportunity for conservation and restoration of dwindling rainforest reserves (Danielsen et al., 2009).

Decoupling agriculture from fossil fuels

Nitrogenous compounds in fertilisers are major contributors to waterway eutrophication and GHG emissions. The Haber-Bosch process for making fertilisers is energy-intensive, consuming 3-5% of the world's natural gas production and releasing large quantities of CO_2 to the atmosphere (Licht et al., 2014). By January 2013, the price of Brent crude oil per barrel rose from around USD 50 to about USD 110. Over the same period, prices for ammonia in Western Europe and the mid-western corn belt in the United States roughly tripled (Chen, 2013).

Several efforts are ongoing in a tantalising research area – creating crop plants that make their own fertiliser. A collaborative project between UK and US scientists aims to design and build a synthetic biological module that could work inside a cell. The project aims to re-engineer the cyanobacterial machinery to fix nitrogen using solar energy as a first step towards transferring the machinery into plants themselves. This has the potential to revolutionise agriculture, and significantly decouple it from the fossil fuels industry. Using synthetic biology, full nitrogen fixation in cereals may be about a decade away (Keasling, 2015). However, partial nitrogen fixation may be available before then (Stokstad, 2016).

Gene editing could usher in a paradigm shift in agricultural biotechnology

As of early 2015, the European Union had granted its member state governments greater power in deciding whether to plant GM crops (BBC News, 2013), which are highly restricted in Europe. The new directive may have split the EU policy landscape, with some states clearly opposed and others in favour (Rabesandratana, 2015). This has come at a time when precision crops are becoming easier to produce through advances in gene editing techniques such as CRISPR/Cas9. Such gene editing techniques are being hailed as low-cost and simple to perform. They are also applicable across all breeding programmes, from large-scale row crops to local and minor species (Von Essen, 2016).

A regulatory shift may bring the application of gene editing techniques to crop production under less scrutiny. Regulators often classify these crops as the product of "new breeding techniques" (NBTs) that are sometimes distinct from classical GM varieties (Nature, 2016). Some mutations edited into the genome already exist in wild plant relatives in nature. The argument then becomes whether the regulator should treat crops produced in this way differently (less stringently) from GM crops.[4] Both the United States and Europe are examining the implications.

More controversial still could be the use of gene editing in domesticated animals. Scientists in Korea and China have created experimental "double-muscled" pigs by editing a single gene, a change that is much less dramatic than those made in conventional, transgenic genetic modification (Cyranoski, 2015). The pigs reportedly provide many of the benefits of the double-muscled cow, such as the Belgian Blue. These include leaner meat and a higher yield of meat per animal – important targets for food security in a bioeconomy. Government agencies may view this more leniently than conventional forms of genetic modification; no "new" DNA has been inserted.

China may be the first to adopt this technology in animals, and this pig could be among the first genetically engineered animals to be approved for human consumption. However, the technique has also been proven to work experimentally in creating double-muscled cows and sheep (Proudfoot et al., 2015).

Policy implications

- A biotechnology revolution in food production has already begun. Production of virtually all animals and crops can benefit from genomics. This includes some critical economic considerations, such as feed efficiency and disease resistance that benefit food security now and into the future. But the benefits are also environmental, although this is even more in its infancy. This message is not well understood at the political level, and the benefits of biotechnology in food production are under-valued and under-reported.

- Governments could better see the advantages of genomics in agriculture, and could more efficiently steer research programmes, by sponsoring programmes that train farmers in genomics. The Irish Beef Data and Genomics Programme is a good example. Gathering relevant field data has been a past limitation, but providing incentives to farmers could help.
- Genomics speeds up breeding, making it more efficient without genetic modification. Campaigns that make this clear could remove public resistance.
- The costs of genomic studies have tumbled through the revolution in next-generation sequencing. Bringing genomics testing to the farms will increase the range of applications as these are clearer to farmers than to researchers. Education and information programmes throughout the agricultural production chain would demystify genomics in agriculture.
- Genetic modification and gene editing have many other applications to offer, but the level of public resistance is greater. However, the United States has cleared GM salmon for human consumption. Its performance in the market may influence consumers in positive or negative ways. Governments could promote the benefits of such GM foods after their safety is guaranteed.

Notes

1. Marker-assisted selection or marker-aided selection (MAS) is a process whereby a marker (morphological, biochemical or one based on DNA/RNA variation) is used for indirect selection of a genetic determinant or determinants of a trait of interest (e.g. productivity, disease resistance, abiotic stress tolerance and quality).
2. An SNP represents a difference in a single DNA building block, called a nucleotide: http://ghr.nlm.nih.gov/handbook/genomicresearch/snp.
3. Over 400 different fatty acids have been identified in seed oils although, remarkably, none have been found to contain the very long chain omega-3 fatty acids.
4. The common white button mushroom (*Agaricus bisporus*) has been modified to resist browning using CRISPR/Cas9, and can be cultivated and sold without further oversight in the United States. The USDA will not regulate a mushroom modified with the gene-editing tool CRISPR/Cas9. The decision means that the mushroom is the first CRISPR-edited organism to be approved by the US government (Waltz, 2016).

References

Alimuddin, Y.G. et al. (2005), "Enhancement of EPA and DHA biosynthesis by over-expression of masu salmon delta-6-desaturase-like gene in zebrafish", *Transgenic Research*, Vol. 14/2, Springer, pp. 159-165.

AquaBounty (n.d.), AquaBounty website, http://aquabounty.com (accessed 7 December 2017).

Aviagen (n.d.), "Research & Development", webpage, http://en.aviagen.com/research-development (accessed 7 December 2017).

Badouin, H. et al. (2017), "The sunflower genome provides insights into oil metabolism, flowering and Asterid evolution", *Nature*, Vol. 546/7656, Nature Publishing Group, London, pp. 248-154.

Baehr, L. (14 July 2014), "This salmon will likely be the first genetically modified animal you eat", Business Insider Gamechangers blog, www.businessinsider.com/this-salmon-will-likely-be-the-first-gmo-animal-you-eat-2014-6?IR=T.

BBC News (2013), "EU changes rules on GM crop cultivation," BBC News, 15 January, www.bbc.com/news/world-europe-30794256.

BDGP (n.d.), "Beef Data and Genomics Programme", webpage, https://www.icbf.com/wp/?page_id=1914 (accessed 12 December 2017).

Bekkevold, D. et al. (2015), "Gene-associated markers can assign origin in a weakly structured fish, Atlantic herring", *ICES Journal of Marine Science*, Vol. 72/6, Oxford University Press, pp. 1790-1801, http://dx.doi.org/10.1093/icesjms/fsu247.

BOLD (n.d.), "Barcode of Life Datasystems", webpage, www.barcodinglife.org (accessed 7 December 2017).

Burt, D.W. (2005), "Chicken genome: Current status and future opportunities", *Genome Research*, Vol. 15/12, Cold Spring Harbor Laboratory Press, United States, pp. 1692-1698.

Cabello, F.C. (2006), "Heavy use of prophylactic antibiotics in aquaculture: A growing problem for human and animal health and for the environment", *Environmental Microbiology*, Vol. 8/7, Wiley Online Library, pp. 1137-1144.

Chen, X.Y. (5 February 2013), "Brent crude oil moves nitrogenous fertilizer prices", Market Realist blog, http://marketrealist.com/2013/02/brent-oil-moves-nitrogenous-fertilizer-prices.

Clark, L.F. (2015), "The current status of DNA barcoding technology for species identification in fish value chains", *Food Policy*, Vol. 54, Elsevier, Amsterdam, pp. 85-94.

Costa, F.O. et al. (2012), "A ranking system for reference libraries of DNA Barcodes: Application to marine fish species from Portugal", *PLoS ONE*, 7/4, Public Library of Science, San Francisco, http://dx.doi.org/10.1371/journal.pone.0035858.

Cunningham, C.O. (2002), "Molecular diagnosis of fish and shellfish diseases: Present status and potential use in disease control", *Aquaculture,* Vol. 206/1-2, Elsevier, Amsterdam, pp. 19-55.

Cyranoski, D. (30 June 2015), "Super-muscly pigs created by small genetic tweak", Nature blog, Vol. 523, Nature Publishing Group, London, pp. 13-14.

Danielsen, F. et al. (2009), "Biofuel plantations on forested lands: Double jeopardy for biodiversity and climate", *Conservation Biology*, Vol. 23/2, Wiley Online Library, pp. 348-358.

Darcé, K. (2010), "A genomics revolution is sweeping agriculture", *San Diego Union-Tribune*, 21 December.

Delgado, C.L. (2003), "Rising consumption of meat and milk in developing countries has created a new food revolution", *The Journal of Nutrition*, Vol. 133/11-2, American Society for Nutrition, Rockville, United States, pp. 3907S-3910S.

D'Hondt, K. et al. (2015), "Reconciling agricultural and industrial needs for an Asian bioeconomy", *Asian Biotechnology and Development Review*, Vol. 17/2, RIS India, New Delhi, pp. 85-130.

FAO (2014), *The State of World Fisheries and Aquaculture 2014 Highlights*, Food and Agriculture Organization of the United Nations, Rome.

FAO (2013), *FAOSTAT* (database), http://faostat.fao.org (accessed 7 December 2017).

FAO (2010), "Species Fact Sheets *Salmo Salar*", webpage, www.fao.org/fishery/species/2929/en (accessed 7 December 2017).

FAO and WHO (2010), *World Review of Fisheries and Aquaculture*, Food and Agriculture Organization of the United Nations, Rome, and World Health Organization, Geneva, www.fao.org/docrep/013/i1820e/i1820e01.pdf.

Gonen, S., et al. (2014), "Linkage maps of the Atlantic salmon (*Salmo salar*) genome derived from RAD sequencing", *BMC Genomics*, Vol. 15, BMC, pp. 166, http://dx.doi.org/10.1186/1471-2164-15-166.

Hanner, R. et al. (2011), "FISH-BOL and seafood identification: Geographically dispersed case studies reveal systemic market substitution across Canada", *Mitochondrial DNA*, Vol. 22-1, Taylor & Francis Online, pp. 106-122.

Hayes, B.J. et al. (2009), "Invited review: Genomic selection in dairy cattle: Progress and challenges", *Journal of Dairy Science*, Vol. 92/2, Elsevier, Amsterdam, pp. 433-443.

Hillier, L.W. et al. (2004), "Sequence and comparative analysis of the chicken genome provide unique perspectives on vertebrate evolution", *Nature*, Vol. 432, Nature Publishing Group, London, pp. 695-716.

Huang, J. and H. Bouis (1996), "Structural changes in the demand for food in Asia", *Food, Agriculture and the Environment Discussion Paper*, No. 11, International Food Policy Research Institute, Washington, DC.

ICISB (n.d.), "Atlantic Salmon Genome Sequence", webpage, www.icisb.org/atlantic-salmon-genome-sequence (accessed 7 December 2017).

IMGC (n.d.), "About us", webpage, http://milkgenomics.org (accessed 7 December 2017).

Jiménez-Sánchez, G. and J.C. Philp (2016), "The bioeconomy, genomics and society", in: *Genomics and Society. Ethical, Legal, Cultural and Socioeconomic Implications*, eds. D. Kumar and R. Chadwick, Academic Press, London.

Jimenez-Sánchez, G. and J. Philp (2015), "Genomics and the bioeconomy: Opportunities to meet global challenges", in *Genomics and Society. Ethical, Legal, Cultural and Socioeconomic Implications*, D. Kumar and R. Chadwick (eds.), Academic Press, London.

Keasling, J. (2015), "Building an opportunity space for synthetic biology", *International Innovation*, Vol. 185, pp. 24-25.

Kondo, M. (2014), "Opportunities for the Australian beef industry in Asia", *Occasional Paper Series*, Number 14/04, Centre for the Study of Rural Australia, Marcus Oldham, Waurn Ponds.

Ledford, H. (23 November 2015), "Salmon approval heralds rethink of transgenic animals. Long-awaited decision by US government authorizes the first genetically engineered animal to be sold as food", Nature blog, Vol. 527, Nature Publishing Group, London, pp. 417-418.

Licht, S. et al. (2014), "Ammonia synthesis by N_2 and steam electrolysis in molten hydroxide suspensions of nanoscale Fe_2O_3", *Science*, Vol. 345/6197, American Association for the Advancement of Science, Washington, DC, pp. 637-640.

Lowenstein, J.H. et al. (2009), "The real *maccoyii*: Identifying tuna sushi with DNA barcodes – contrasting characteristic attributes and genetic distances", *PLoS ONE*, 4/11, Public Library of Science, San Francisco, http://dx.doi.org/10.1371/journal.pone.0007866.

Ly, D. et al. (2013), "Relatedness and genotype-by-environment interaction affect prediction accuracies in genomic selection: A study in Cassava", *Crop Science*, Vol. 53, Crops Science Society of America, Madison, pp. 1312-1325.

Marco, M.L. and M.H.J. Wells-Bennik (2008), "Impact of bacterial genomics on determining quality and safety in the dairy production chain", *International Dairy Journal*, Vol. 18, Elsevier, Amsterdam, pp. 486-495.

McAndrew, B. and J. Napier (2011), "Application of genetics and genomics to aquaculture development: Current and future directions", *The Journal of Agricultural Science*, Vol. 149, Canadian Center for Science and Education, Toronto, pp. 143-151.

McClure, M.C. et al. (2013), "Imputation of microsatellite alleles from dense SNP genotypes for parentage verification across multiple *Bos taurus* and *Bos indicus* breeds", *Frontiers in Genetics*, Vol. 4, Frontiers, Lausanne, pp. 176, http://dx.doi.org/10.3389/fgene.2013.00176.

McKechnie, I. et al. (2014), "Archaeological data provide alternative hypotheses on Pacific herring (*Clupea pallasii*) distribution, abundance, and variability", *Proceedings of the National Academy of Sciences*, Vol. 111/9, pp. E807–E816.

Milne, R. (2015), "Norway fund excludes four Asian groups over palm oil links", *The Financial Times,* 17 August, www.ft.com/intl/cms/s/0/b8305bbe-44c7-11e5-b3b2-1672f710807b.html#axzz3IKPEGZN4.

Moustafa, K. (2016), "Toward future photovoltaic-based agriculture in sea", *Trends in Biotechnology*, Vol. 34/4, Elsevier, Amsterdam, pp. 257-259.

Nature (12 April 2016), "Breeding controls. Scientists must help to inform regulators wrestling with how to handle the next generation of genetically engineered crops", Nature blog, Vol. 532, Nature Publishing Group, London, pp. 147.

OECD (2015), "Genomics for sustainable development in emerging economies: Food, environment, and industry", *Workshop Report*, 18-19 May 2005, OECD, Paris, www.oecd.org/officialdocuments/publicdisplaydocumentpdf/?cote=DSTI/STP/BNCT(2015)6&docLanguage=En.

Overholtz, W.J. et al. (2008), "An ecosystem approach for assessment advice and biological reference points for the Gulf of Maine–Georges Bank Atlantic herring complex", *North American Journal of Fisheries Management*, Vol. 28/1, Taylor & Francis Online, pp. 247-257.

Palaiokostas, C. et al. (2013), "Mapping and validation of the major sex-determining region in Nile Tilapia (*Oreochromis niloticus* L.) using RAD sequencing", *PLoS ONE*, Vol. 8/7, Public Library of Science, San Francisco, http://dx.doi.org/10.1371/journal.pone.0068389.

Palenzuela, P. et al. (2015), "Large-scale solar desalination by combination with CSP: Techno-economic analysis of different options for the Mediterranean Sea and the Arabian Gulf", *Desalination*, Vol. 366, Elsevier, Amsterdam, pp. 130-138.

Pollak, E.J. et al. (2012), "Genomics and the global beef cattle industry", *Animal Production Science*, Vol. 52/3, CSIRO Publishing, Clayton, Australia, pp. 92-99.

Proudfoot, C. et al. (2015), "Genome edited sheep and cattle", *Transgenic Research*, Vol. 24/1, Springer, pp. 147-153.

Rabesandratana, T. (2015), "Europe's rifts over transgenic crops deepen at key deadline. Policy differences among 28 EU member states come into sharper focus in the wake of major reform", *Science*, Vol. 350, American Association for the Advancement of Science, Washington, DC, pp. 18-19.

Raven, L.A. et al. (2013), "Genes of the RNASE5 pathway contain SNP associated with milk production traits in dairy cattle", *Genetics Selection Evolution*, Vol. 45/1, Springer, pp. 25, http://dx.doi.org/10.1186/1297-9686-45-25.

Science Daily (11 June 2014), "Milestone in salmon research: Genome fully sequences," Science News blog, www.sciencedaily.com/releases/2014/06/140611093238.htm.

Singh, R. et al. (2013a), "Oil palm genome sequence reveals divergence of inter-fertile species in old and new worlds", Nature Letter, Vol. 500/7462, Nature Publishing Group, London, pp. 335-339.

Singh, R. et al. (2013b), "The oil palm SHELL gene controls oil yield and encodes a homologue of SEEDSTICK", *Nature*, Vol. 500/7462, Nature Publishing Group, London, pp. 340-344.

Sommerset, I. et al. (2005), "Vaccines for fish in aquaculture", *Expert Review of Vaccines*, Vol. 4/1, Taylor & Francis Online, pp. 89-101.

Stokstad, E. (2016), "The nitrogen fix", *Science*, Vol. 353.6305, American Association for the Advancement of Science, Washington, DC, pp. 1225-1227.

Tarr, P. (24 July 2013), "Full genome map of oil palm indicates a way to raise yields and protect rainforest", Cold Spring Harbor Laboratory News blog.

Technology Strategy Board (2010), "Boosting global food security", October, https://connect.innovateuk.org/documents/3285671/6079410/Boosting+Global+Food+Security.pdf/b0ee6ed1-2e98-41f2-b4e2-2b594db4a035.

The Economist (2015), "Something fishy. How to make farmed salmon into vegetarians", *The Economist*, 11 July, https://www.economist.com/news/science-and-technology/21657354-how-make-farmed-salmon-vegetarians-something-fishy.

Tinline, J. (2015), "Can farmed fish meet the world's appetite for seafood", BBC News, 12 June, www.bbc.com/news/business-33068446.

University of Glasgow (2015), "Selective breeding can produce salmon resistant to sea lice", University News, 18 August, www.gla.ac.uk/news/headline_416474_en.html.

USDA Foreign Agriculture Service (2012), "Malaysia: Stagnating palm oil yields impede growth", *Commodity Intelligence Report*, United Nations Department of Agriculture, 11 December.

Usher, S. et al. (2015), "Field trial evaluation of the accumulation of omega-3 long chain polyunsaturated fatty acids in transgenic *Camelina sativa*: Making fish oil substitutes in plants", *Metabolic Engineering Communications*, Vol. 2, Elsevier, Amsterdam, pp. 93-98.

van der Zee (2017), "UK is 30-40 years away from 'eradication of soil fertility', warns Gove", *The Guardian*, 24 October, https://www.theguardian.com/environment/2017/oct/24/uk-30-40-years-away-eradication-soil-fertility-warns-michael-gove.

Von Essen, G. (2016), "Precision breeding – smart rules for new techniques", *European Biotechnology*, Vol. 15, Biocom AG, Berlin, pp. 3.

Waltz, E. (14 April 2016), "Gene-edited CRISPR mushroom escapes US regulation. A fungus engineered with the CRISPR–Cas9 technique can be cultivated and sold without further oversight", Nature blog, Vol. 532, Nature Publishing Group, London, pp. 293.

Waughrey, D. (5 June 2017), "Tuna 2020 traceability declaration: Stopping illegal tuna from coming to market", World Economic Forum blog, www.weforum.org/agenda/2017/06/tuna-2020-traceability-declaration-stopping-illegal-tuna-from-coming-to-market.

World Fishing & Aquaculture (2014), "Aquagen identifies IPN resistance gene", 20 June, www.worldfishing.net/news101/industry-news/aquagen-identify-ipn-resistance-gene (accessed 31 January 2018).

Part II.

Biorefinery models and policy

Chapter 5.

What is a biorefinery: Definitions, classification and general models

This chapter explores biorefinery models and their status, setting the stage for later chapters that focus more on public policy. Biorefinery models have evolved according to needs from the first ethanol mills using food crops as feedstocks to more complex (and more expensive) models using feedstocks other than food crops. The ultimate goal is the widespread application of the integrated biorefinery that can use multiple feedstocks and generate multiple products (fuels, chemicals, materials, electricity). However, these are still not ready for the market and are seen as high-risk investments. Building the first-of-kind flagship plants is proving difficult. Meanwhile, marine biorefineries, which offer similar advantages, remain difficult to design and build. And other yet more novel biorefinery concepts are arising.

Introduction

Definitions of a biorefinery are important for gathering data, observing trends and investing public funds. It is necessary then, to identify what actually happens in a generalised model of a biorefinery and then explore the different models and definitions. Figure 5.1 is a schematic of general processes and the order in which they occur.

Figure 5.1. **General schematic of a biorefinery**

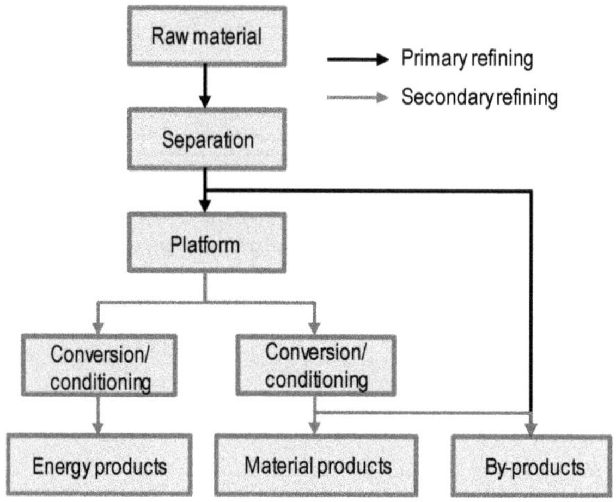

Source: Peters (2011), "The German biorefinery roadmap".

Box 5.1. **Examples of definitions of biorefinery**

The term "green biorefinery" has been defined as "complex systems based on ecological technology for comprehensive (holistic), material and energy utilization of renewable resources and natural materials using green and waste biomass and focalising on sustainable regional land utilization". The term "complex systems" can now be regarded as "totally integrated systems" (Kamm et al., 1998).

According to Kamm et al. (2007, 2006), the US Department of Energy (DOE) uses the following definition: "A biorefinery is an overall concept of a processing plant where biomass feedstocks are converted and extracted into a spectrum of valuable products. Its operation is similar to that of petrochemical refineries".

The US National Renewable Energy Laboratory (NREL) uses the following definition: "A biorefinery is a facility that integrates biomass conversion processes and equipment to produce fuels, power and chemicals from biomass. The biorefinery concept is analogous to today's petroleum refineries, which produce multiple fuels and products from petroleum. Industrial biorefineries have been identified as the most promising route to the creation of a new domestic biobased industry".

The International Energy Agency (IEA) describes the biorefinery as "the sustainable processing of biomass into a spectrum of marketable products (food, feed, materials, chemicals) and energy (fuels, power, heat)". This means that biorefinery can be a concept, a facility, a process, a plant or even a cluster of facilities.

A future definition of biorefinery could include processes that use living organisms to convert waste products from non-biogenic sources, including CO_2 from fossil fuel combustion.

Sources: Kamm, B. et al. (2007), "Biorefineries – industrial processes and products"; Kamm, B. et al. (2006), "Biorefinery systems – an overview"; Kamm, B. et al. (eds.) (1998), "Die grüne Bioraffinerie"; Schieb, P.-A. et al. (2015), *Biorefinery 2030: Future Prospects for the Bioeconomy*, http://dx.doi.org/10.1007/978-3-662-47374-0.

The International Energy Agency (IEA Bioenergy Task 42 Biorefinery, 2012) described a biorefinery as "the sustainable processing of biomass into a spectrum of marketable products (food, feed, materials, chemicals) and energy (fuels, power, heat)". This definition suggests that biorefineries should produce both non-energetic and energetic outputs, and applies to product-driven biorefinery processes. Both primary products and energy-driven processes are considered as true biorefinery approaches provided the final goal is the sustainable processing of biomass (de Jong and Jungmeier, 2015). Some existing definitions of "biorefinery" are shown in Box 5.1.

The IEA biorefinery classification system is useful in clarifying different models in operation and under development. The widespread adoption of the IEA system would help clarify several issues. Its classification approach consists of four main features that can identify, classify and describe the different biorefinery systems: platforms, products (energy and bio-based materials and chemicals), feedstocks and conversion processes.

The raw material or feedstock has a highly varied range of organic materials (containing carbon).[1] Feedstocks can be grouped. Energy crops from agriculture (e.g. starch crops, short rotation forestry) constitute the major feedstocks. Biomass residues from agriculture, forestry, trade and industry (e.g. straw, bark, used cooking oils, waste streams from biomass processing) form another major category; these biorefineries are the most promising for future progress. Even less conventional feedstocks include municipal solid waste (MSW) and waste industrial gases such as CO and H_2 from the steel-making process.

Figure 5.2 **Conversion processes in a fuel biorefinery**

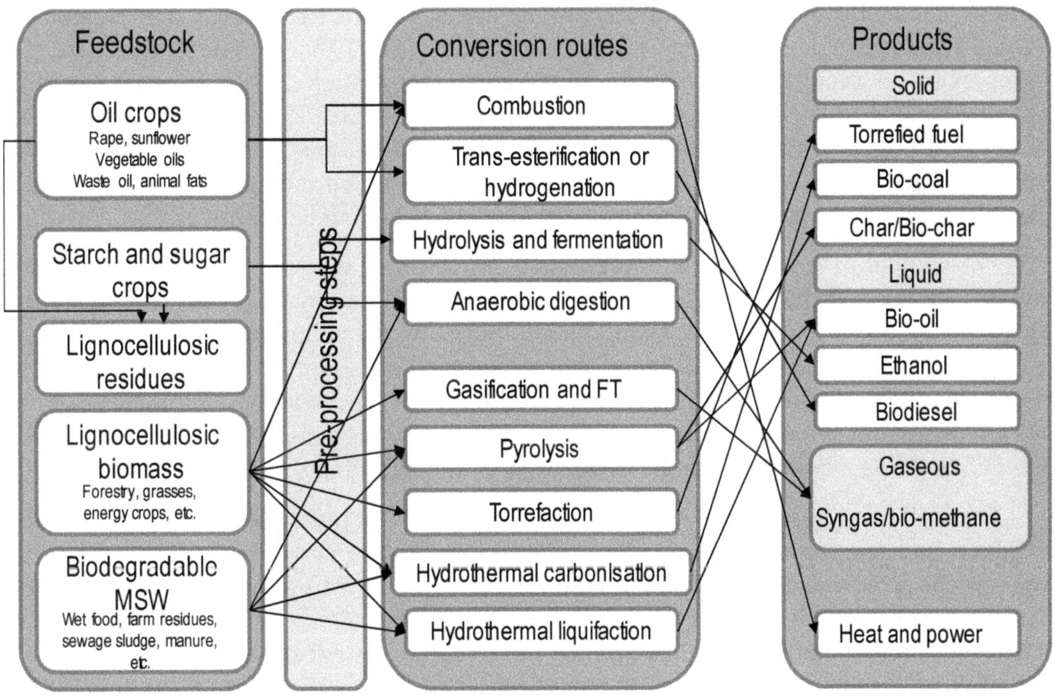

Note: MSW = municipal solid waste.

Concerning conversion processes, the IEA classification system identifies four main groups: biochemical (e.g. fermentation, enzymatic conversion); thermochemical (e.g. gasification, pyrolysis); chemical (e.g. acid hydrolysis, steam explosion, esterification);

and mechanical processes (e.g. fractionation, pressing, size reduction).Energy products can be usually considered as liquid fuels (e.g. ethanol, biodiesel, bio-based jet fuel), but biogas is also possible. Wood chips, pellets and lignin are possible solid fuel outputs. Material products could be any of a large number of bio-based chemicals, plastics and textiles. Energy products might also be residues at the end of the process that can be burned to generate electricity and/or heat. By-products could include animal feed and soil conditioners.

The diversity of biorefineries is already large, and Figure 5.2, showing different feedstocks and conversion processes, testifies to this diversity.

The IEA Biorefinery Complexity Index

All current models of biorefineries are high-risk investments, and even Brazilian ethanol mills have gone through difficult times. The industry has had a spate of bankruptcies, a problem caused by the global financial crisis, adverse weather and low sugar prices (Soybean and Corn Advisor, 2015). Biorefineries range from single feedstock-single product operations to multiple feedstock-multiple products. In other words, the complexity of biorefineries varies greatly. Arguably, prospects for economic viability mount in tandem with complexity. When conditions dictate, one feedstock can be replaced by another, and the product stream can be changed.

However, different degrees of complexity make it challenging for industry, decision makers and investors to identify the most promising short-, medium- and long-term options, including their technological and economic risks. In response, IEA Bioenergy Task 42 published a Biorefinery Complexity Index (BCI) that can help calculate the complexity of some selected biorefinery concepts (Jungmeier, 2014). It bears a strong resemblance to the Nelson Index used in petro-refineries. The Nelson Index is an indicator for the investment intensity, cost index and value addition potential of the refinery, and the refinery's ability to process feedstocks into value-added products (Wikipedia, n.d.). As the refinery becomes more complex, it is also considered to be more flexible.

The number of features in the biorefinery – used to calculate the "Feature Complexity Index" (FCI) – and the "Technology Readiness Level" (TRL) make up the essence of the BCI calculation. With each new feature in a biorefinery, the complexity increases. A high TRL of a feature has lower technical and economic risks, and so a lower complexity. Thus, the number of features determines the complexity of a commercial application, in which all features are commercially available. Conversely, in non-commercial applications, the FCI and TRLs both increase the complexity of the biorefinery system.

The TRL can assess each of a biorefinery's four features (platforms, feedstocks, products and processes) using standard descriptions from 1 (lowest) to 9 (highest, with the system proven and ready for full commercial deployment (see also Chapter 8). The feature complexity (FC) for each single feature of a biorefinery is calculated based on the TRL. Once the number of features and the FC of each single feature are known, the FCI for each of the four features can be calculated. The BCI is the sum of the four FCIs. The Biorefinery Complexity Profile (BCP) is introduced to simplify the presentation. Jungmeier (2014) provides details of how to make the appropriate calculations.

Equation 5.1. The biorefinery complexity profile

$$BCP = BCI \times (FCI_{Platforms}/FCI_{Feedstocks}/FCI_{Products}/FCI_{Processes})$$

The BCP and the BCI arguably can compare different biorefinery concepts and their development potential. As the BCI increases, the biorefinery moves increasingly beyond "state of the art". Furthermore, this system is flexible as it can consider changes in TRL of features through research and development. It can therefore help address the economic and technical risks for any given biorefinery project or concept. This, in turn, will help public- and private-sector investors make decisions.

Biorefinery types: A brief description

There are myriad different concepts arising for different biorefineries. However, these are concentrated into a small number of what can be seen as biorefinery "types". Many are described in the literature, such as The Biorefinery Roadmap (2012) (Federal Government of Germany, 2012). Excellent unit process descriptions are found in the Star-COLIBRI (2011) European Biorefinery Joint Strategic Research Roadmap. The former is especially helpful (Figure 5.3) as it also estimated the status of technological development at the time of publication. Things have moved on since 2012, but this status has not really changed significantly. Changes, too, can be described.

Figure 5.3. **Development status of various biorefinery models**

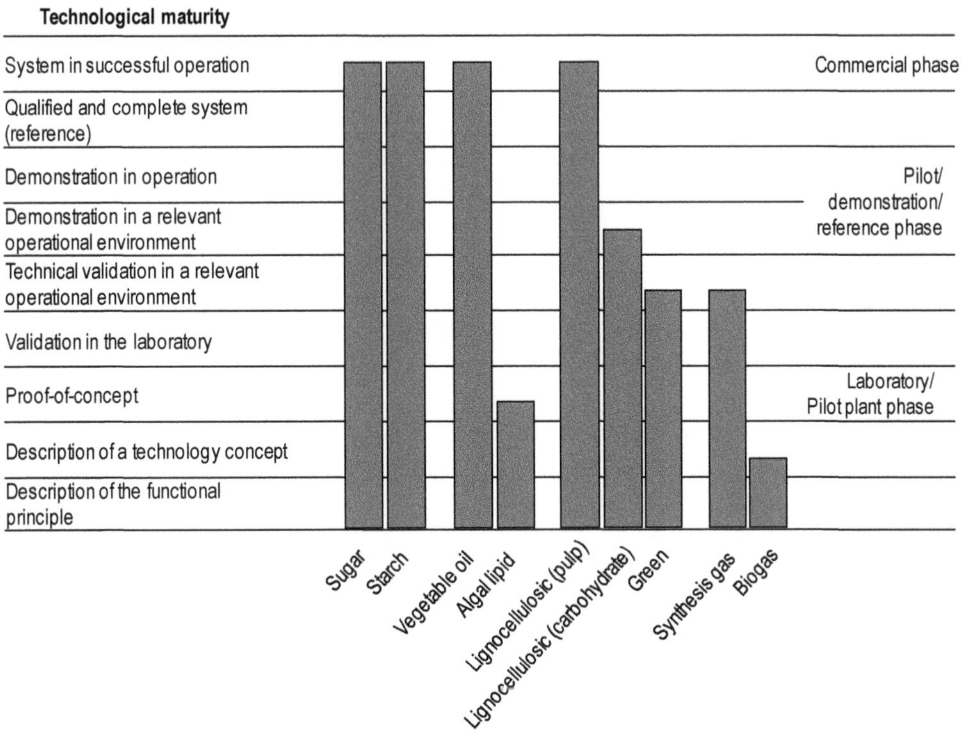

Source: Federal Government of Germany (2012), "Biorefineries roadmap".

The typical sugar cane biorefinery

In terms of economic sustainability, Brazilian sugar cane is the most favoured feedstock for biorefineries at present (e.g. UK Bioenergy Strategy, 2012). As of 2011, Brazil had 490 sugar cane ethanol plants and biodiesel plants (BRBIOTECH-CEBRAP, 2011). As of mid-2016, for various reasons, this number had declined to somewhere around 300 operational sugar cane/ethanol plants, with some of the closures permanent. Figure 5.4 shows the first-generation bioethanol production process from sugar cane. The typical Brazilian ethanol mill has a processing capacity of 500 tonnes of sugar cane per hour (wet basis), equivalent to 2 million tonnes per year. At the industrial level, most sugar cane in Brazil is processed through an integrated production chain, allowing sugar production, industrial ethanol processing and electricity generation from by-products. The typical steps for large-scale, highly optimised production of sugar and ethanol include milling, electricity generation, fermentation, distillation of ethanol and dehydration.

Figure 5.4. **Integrated first- and second-generation ethanol production from sugar cane**

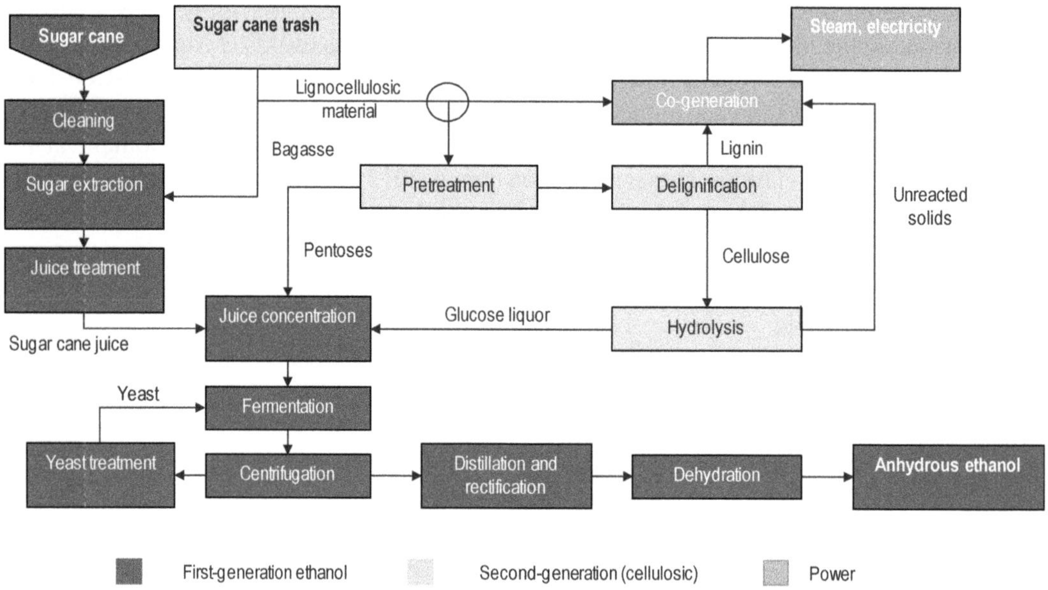

Source: Dias et al. (2013), "Biorefineries for the production of first- and second-generation ethanol and electricity from sugar cane".

In the Brazilian sugar cane industry, large amounts of lignocellulosic materials, especially bagasse, are readily available, typically as by-products of sugar and ethanol. Most of the bagasse produced in the mills, where sugar cane juice is separated from the fibre, supplies energy for the bioethanol production process in cogeneration systems. It is commercially and technically feasible in Brazil to sell sugar cane lignocellulosic fractions to the grid as fuels in electricity production (Cardona et al., 2010). If electricity prices are favourable, more lignocellulosic material may be diverted for production of steam and electricity (see the circle on Figure 5.4). The opposite occurs when ethanol prices are more attractive (Dias et al., 2013).

Lignocellulosic or cellulosic biorefinery

Lignocellulose is composed of carbohydrate polymers (cellulose, hemicellulose), and an aromatic polymer (lignin). It is the most abundant raw material for biorefining as it contains large amounts of fermentable sugars. However, the sugars needed for fermentation are tightly bonded within the lignocellulose. This becomes a barrier to using lignocellulose from biomass in biorefining. Much of the technical effort to unleash this vast bounty for biorefining is related to overcoming this recalcitrance of the feedstock; the "conversion" has been the bottleneck.

Lignocellulosic biomass can be grouped into four main categories (Tan, Yu and Shang, 2011):

1. agricultural residues (e.g. corn stover and sugar cane bagasse)
2. dedicated energy crops
3. wood residues (including sawmill and paper mill discards)
4. municipal paper waste.

Costs vary between types of plants. Second-generation biofuel plants may have capital costs five times greater than starch ethanol plants (Wright and Brown, 2007). For first-generation bioethanol, the most significant cost was feedstock (Carriquiry et al., 2011). About 40-60% of the total operating cost of a typical biorefinery is related to the feedstocks chosen (Parajuli et al., 2015). However, the most significant cost for second-generation cellulosic biofuels may be conversion of woody biomass into fermentable sugars.

A crisis of sorts has arrived in cellulosic biorefining. Through its Renewable Fuels Standard (RFS), the US Environmental Protection Agency (EPA) is enforcing 230 million gallons of cellulosic biofuel blending for 2016. The RFS statute, however, nominally requires 4.25 billion gallons, which represents a 95% reduction. Technical problems surrounding conversion have proven so intractable that only a handful of these biorefineries have become commercially viable (Figure 5.5).

Figure 5.5. **Global capacity in cellulosic biorefining**

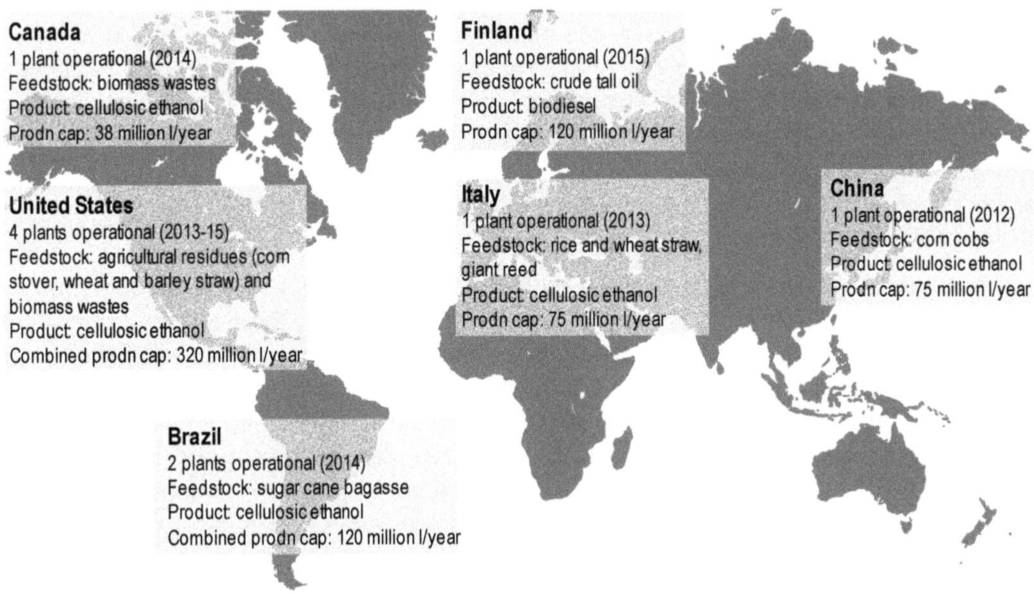

Note: prodn cap = production capacity; l = litre.

As a result of these technical barriers and policy uncertainty, investments in these biorefineries have been drastically reduced. A new one was approved for construction in July 2016 in Renfrew, Ontario, Canada. However, it may be the only commercial-scale cellulosic biofuel project that has gained approval anywhere in the world over the past two years. Financing for the Renfrew plant is overwhelmingly from the public sector. It will convert forestry waste into Ensyn biocrude for further processing in oil refineries. The California Air Resources Board granted key regulatory approvals to Ensyn pursuant to the low-carbon fuel standard on the company's application for its biocrude renewable fuel oil. California oil refineries will use it in co-processing (Biodiesel Magazine, 2016).

Waste biorefineries: Rubbish to bio-based products and electricity

Although they can be categorised as lignocellulosic biorefineries, domestic waste biorefineries are treated separately here to highlight their future potential. The use of domestic waste materials as feedstock for biorefineries promises to be the most sustainable approach, provided waste materials are collected efficiently. A large amount of waste is available for feedstock, but political will is needed to create incentives for its collection.

Using municipal waste not only reduces the amount of waste going to landfills, it also breaks the link between food crops and bioethanol production. At full production, the waste biorefinery in Vero Beach, Florida, for example, (INEOS, 2013) is expected to produce 8 million gallons of advanced cellulosic bioethanol and 6 megawatts (gross) of renewable power. It uses renewable biomass including yard, vegetative and agricultural wastes. The waste material goes through a gasification process to create synthesis gas (syngas). Syngas can then be used to manufacture a range of chemicals, either through synthetic chemistry or fermentation (Latif et al., 2014). The heat recovered from the hot syngas is fed into a steam turbine and used to generate renewable electricity. The renewable electricity powers the facility; the excess electricity is expected to power as many as 1 400 homes in the Vero Beach community. A relatively small facility, it has 60 full-time employees and provides USD 4 million annually in payroll to the local community.

The Vero Beach project is also a good example of a public-private partnership (PPP). These are deemed to be a way to get high-risk biorefineries built in the absence of substantial interest from venture capital investors. Ineos Bio and New Planet Energy, Florida, in a PPP with the US Department of Energy (USD 50 million cost-matched grant) and the US Department of Agriculture (USD 75 million loan guarantee), have constructed this waste biorefinery.

Algal biorefineries

Both micro- and macroalgae are extremely promising feedstocks for future biorefineries for a variety of reasons (IEA Bioenergy Task 39, 2011). First, the land requirement for algae is much less than for terrestrial crops, thus alleviating pressure on food crops. Second, they grow rapidly and have a higher solar conversion efficiency than most terrestrial plants. Third, they can be harvested batch-wise or continuously almost all year-round. Fourth, they can use waste CO_2 sources, thereby potentially mitigating the release of GHGs into the atmosphere. Finally, they could generate a vast amount of oil compared to terrestrial crops (Table 5.1); the differences are of magnitude orders.

However, of all the road transport biofuels reviewed by Accenture (2009), algal technology was deemed to be the most difficult and will take the longest to achieve commercial scale. Nonetheless, some companies claim that the first commercial plants will be available soon in various parts of the world, including Australia, Europe, the Middle East, New Zealand and the United States (Pienkos and Darzins, 2009). That prediction of 2009 remains accurate, as marine biorefining still presents large technical challenges. The design and engineering principles for marine biorefining are in their infancy compared to biorefineries for terrestrial crops. The development of stable cultivation technologies – harvesting, product extraction and biorefinery processes – represent the main challenges of algal biotechnology for production of high-value or bulk products.

Genetic engineering for strain improvement and higher product yields, and the need to gain market and regulatory acceptance of such organisms, are other major challenges (Sayre et al., 2013).

Table 5.1. **Oil yields from various terrestrial plants compared to algae**

Crop	Oil yield (gallons/acre)
Corn	18
Cotton	35
Soybean	48
Mustard seed	61
Sunflower	102
Rapeseed	127
Jatropha	202
Oil palm	635
Algae	10 000

Source: Pienkos (2009), "The promise and challenges of microalgal-derived biofuels".

The seaweed (macroalgae) industry is small but mature, and has plenty of scope for expansion. Nearly 7.5-8 million tonnes of wet seaweeds are harvested worldwide per year (Subba Rao and Mantri, 2006), but the treatment of spent seaweed is challenging. Apart from the oil, macroalgae contain various higher-value chemicals, such as plant proteins, alginates and phenolics. Moreover, fermentation of seaweed hydrolysates can produce many by-products, such as glycerol, organic acids (e.g. acetate, succinate), biomass protein and other minor products (Wei et al., 2013). And because seaweed biomass does not contain lignin, residuals after fermentation can be used as animal feed or a feed supplement. Therefore, there is great scope for cascading use of biomass with algae and cyanobacteria (Ducat et al., 2011). For example, a study has examined the production of ethanol from spent biomass generated from the seaweed processing industry using baker's yeast. The process has potential for converting galactose and alginate monomers to bioethanol through fermentation (Sudhakar et al., 2016).

Certain caveats must be invoked when discussing the potential of algal technologies, especially microalgal technologies. Several comprehensive analyses study the design and economics of microalgal processes, but they leave the actual species undetermined. By doing so, the assumptions of the analyses may be inaccurate. With this in mind, the need for rapid, accurate and defendable taxonomic identification of microalgae and cyanobacteria strains is paramount for culture collections, industry and academia, particularly when addressing issues of intellectual property and biosecurity (Emami et al., 2015).

Similarly, there are locations with sufficient year-round levels of sunlight close to plenty of water. Further, they are not far from carbon-intensive industries that can supply inexpensive CO_2. And they have access to developed road and rail networks that can support distribution of raw materials and end products. But these locations are by no means commonplace (Klein-Marcuschamer et al., 2013).

The National Marine Bioenergy Research Centre, in collaboration with the department of Biological Engineering of the Inha University at Incheon, Korea, has tested an experimental algae production system. Algae are produced in semi-permeable membranes in the sea. In this system, no energy for the culturing needs to be added as the sea movement keeps the culture moving. Further, as seawater contains more nutrients than fresh water, no extra nutrients need to be added; they are taken up through the semi-permeable membrane.

The experimental set up produced bioethanol up to three times higher from red or brown algae than from sugar beet or sugar cane, the best performing land energy crops. For production of biodiesel, the yield was even up to ten times higher from microalgae than from oil palm, which is the best performing biodiesel production crop on land. This production system has, in fact, passed all government criteria. The oil produced has better quality than palm biodiesel.

Waste gas and syngas biorefineries

Gas fermentation offers an opportunity to use resources as diverse as industrial waste gases, coal and municipal solid waste (after gasification) to produce fuels and chemicals. A 1995 demonstration at the laboratory scale showed the feasibility of converting gases to bioplastics (Tanaka et al., 1995). Hydrogen, oxygen and CO_2 were converted to a bio-based, biodegradable plastic in the absence of another source of carbon. Other bio-based products have been shown to be feasible at laboratory scale. For example, the steel mill off-gas CO and syngas can be fermented into a variety of useful products, such as ethanol and 2,3-butanediol (Köpke et al., 2011).

However, taking gas fermentation technology to commercialisation has taken a long time. LanzaTech, a waste gas-to-fuel and -chemicals start-up founded in New Zealand, converted steel mill waste gases to ethanol at demonstrator level in 2013. It produced roughly 380 cubic metres (m^3) of ethanol per year at a steel mill near Shanghai in the People's Republic of China (hereafter "China") (Pavanan et al., 2013). In 2014, the company closed a USD 60 million investment from the New Zealand Superannuation Fund, a sovereign wealth fund, to develop the technology further.

A system to be built at an ArcelorMittal steel mill in Ghent, Belgium would be about 30 times larger than the Shanghai plant, producing some 47 000 tonnes of ethanol a year (Clark, 2015). It will cost EUR 87 million to install, and the project has received EUR 10 million in EU research funding. If the system at Ghent proves to be commercially viable, ArcelorMittal, the world's largest steel maker, hopes to install it across its operations. This move could eventually produce up to 10% of Europe's bioethanol a year.

The steel industry has long struggled to deal with its emissions (OECD, 2015). The top three industrial GHG emitters are steel, cement and chemicals. This biorefining technology would help steel makers reduce emissions, and also add value to their core business. It also does not compete for land or interfere with food as no crops are required.

The integrated chemical and biological biorefinery concept

The integrated biorefinery (Figure 5.6) would make full use of all the components of multiple feedstocks (particularly cellulosic). It would produce value-added multiple co-products including energy (electricity and steam) and various bio-based chemicals and plastics, along with fuel-grade ethanol or other fuels. It might even be able to create other products such as paper.

In this concept, chemicals and fuel production are integrated within a single operation where high-value products become an economic driver. These products provide higher margins to support low-value fuel, leading to a profitable biorefinery operation that also exhibits an energy impact. This is how many petrochemical oil refineries operate – the 7-8% of crude oil for chemical production results in 25-35% of the annual profits of integrated petrochemical refineries (Bozell, 2008). Many configurations are possible depending on the choice of chemicals to be manufactured on-site.

Figure 5.6. **Schematic of a generalised integrated biorefinery**

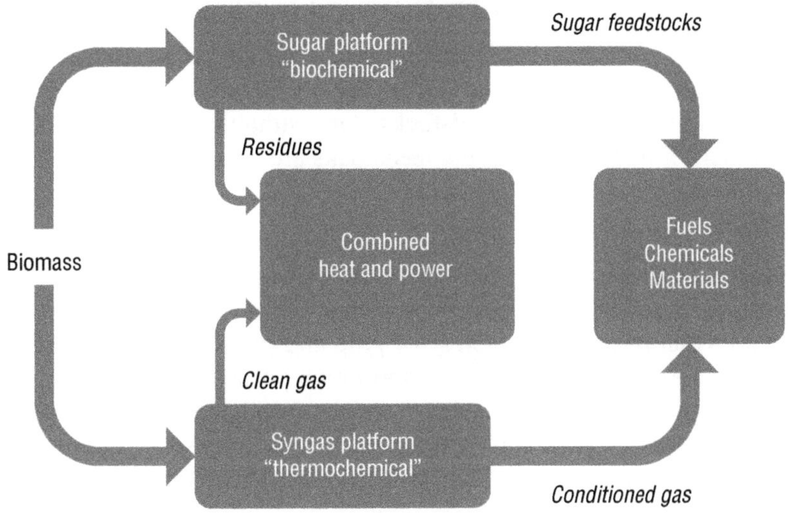

Source: OECD (2017), *The Next Production Revolution*, http://dx.doi.org/10.1787/9789264271036-en.

Such a biorefinery is obviously technically complex, even more so than a petrochemical facility. But it has some advantages compared to single feedstock, single product biorefineries that make this model particularly attractive. First, it can switch between feedstocks and products when, for example, one particular feedstock is too expensive; switching between feedstocks helps cope with seasonal availability (Giuliano et al., 2016). Integration avoids the low-margins trap of producing high-volume fuels (OECD, 2014). Specifically, there is less fractional market displacement required for cost-effective production of high-value co-products as a result of the economies of scale provided by the primary product (Lynd et al., 2005); the economies of scale provided by a full-size biorefinery lower the processing costs of low-volume, high-value co-products. In addition, biorefineries maximise value generated from heterogeneous feedstock, making use of component fractions. Common process elements are involved, lowering the need for equipment duplication, with subsequent decreases in capital cost. Co-production can provide process integration benefits (e.g. meeting process energy requirements with electricity and steam co-generated from process residues). Finally, it can operate like a "waste exchange".

A lesson can be observed from US biodiesel production from soybean oil. Over 2005-08, the price of soybeans doubled. Many biodiesel production plants halted production, delaying construction of new plants (Starkey, 2008). Such issues may be avoidable if low production volume, higher value-added products can also be made at the same site. The integrated biorefinery also gives the flexibility to use different feedstocks if one feedstock is unavailable.

The benefits notwithstanding, several defining challenges are proving difficult to overcome (Cheali et al., 2015). For example, it is difficult to achieve maximum efficiency with improved designs or to expand by integration of conversion platforms or upstream and downstream processes. Other challenges relate to accounting for a wide range of feedstock, processing paths and product portfolios (Tsakalova et al., 2015). Whereas fossil fuel-based processes (i.e. local supply and value chains) formulate local/regional solutions, biorefineries develop solutions on a global basis. Finally, design challenges relate to feedstock characteristics, feedstock quality and availability; trade-offs between energy consumption for feedstock and product distribution, production and product market prices).

Real examples of truly integrated biorefineries are not yet available. This is not the result of low oil and gas prices, but rather due to technical challenges in perfecting processes with waste materials as feedstocks. One suitable candidate is the ARD-BRI complex in northern France, although the feedstocks are food crops (Box 5.2).

Wood biorefineries

Again, there is much cross-over between cellulosic and integrated biorefineries. Some issues are identical, especially the conversion technologies. Wood biorefining makes sense in many countries that have a long history of pulp and paper-making. The relatively high energy density of wood is attractive for transportation purposes. An advantage enjoyed by pulp and paper mills in biorefining stems from the perfected kraft processing of wood. The kraft process converts wood into wood pulp, which consists of almost pure cellulose fibres, the main component of paper. The process treats wood chips with a hot mixture of water, sodium hydroxide and sodium sulphide, known as white liquor; this breaks the bonds that link lignin, hemicellulose and cellulose.

Unlike cellulosic biorefineries, wood contains much more lignin than, say, agricultural residues. Lignin is challenging to biorefine, but remains a major potential source of all manner of aromatic compounds. Aromatics are extremely important industrial chemicals, and bio-based drop-ins or alternatives are not easy to produce. The interest in lignin as a source of chemicals or materials is increasing; processes for lignin isolation from kraft processes are being installed.

The potential of lignin is not just in drop-in alternative chemicals; it is a polymer that can be derivatised for various applications. Lignin epoxide, for example, can be used for printed circuit board, segmented polyurethane plastics and others. The new wood biorefinery processes will produce sulphur-free lignin, which offers several advantages in chemical and material production. Still, despite these advantages, lignin sulphonates and lignin sulphates from "old" pulping processes exhibit performance properties because of the sulphonate and sulphate groups.

Lignin applications are becoming increasingly visible: the amounts of lignin produced annually are huge. The variety of valuable compounds that could be produced from the aromatic lignin could answer doubts over the ability of bio-production to produce aromatics.

Several other future options include: extraction of cellulose fibre and valuable products from bark (e.g. fine chemicals and pharmaceuticals), wood extractives (fatty acids used in products like water-based resins), pulping liquor (carbohydrates used as hydrocolloids, emulsifiers and food ingredients). There are several comprehensive sources of information on the chemistry of wood (e.g. Sjostrom, 1993).

Bioökonomierat (2016) has suggested two major lines of development for innovative wood biorefinery processes that concur with the above analysis:

- digestion of wood with subsequent enzymatic hydrolysis to obtain fermentation feedstocks and lignin
- thermochemical processes that provide fuel or basic chemicals as a result of pyrolysis or gasification.

Box 5.2. The Agro-industrie recherches et développements biorefinery hub and Bioraffinerie recherches et innovation at Bazancourt-Pomacle, northern France

Agro-industrie recherches et développements (ARD) is a mutualised private research structure, owned by major players in the French agri-business as well as regional farming co-operatives, the latter being a particular strength. It was created in 1989 by exploiting the notion of value creation through non-food applications to find new opportunities from the produce of its shareholders (e.g. cereals, sugar beet, alfalfa, oilseeds). Subsequently, ARD started two subsidiaries – Soliance (molecules for cosmetic products) and BIODEMO, the largest capacity demonstration platform in France, which has hosted Amyris, BioAmber and Global Bioenergies.

The innovation hub Bioraffinerie recherches et innovation (BRI) is an open hub in the field of biorefining. BRI brings together various biorefineries at Bazancourt-Pomacle, the R&D centre ARD, as well as the French engineering schools École centrale Paris, Agro Paris Tech and NEOMA Business School. Therefore, it covers the value chain from fundamental research to the pre-industrial prototype.

It has had public financial support from the Ministry of Industry of France, the General Council of the Marne Département, the Region Champagne-Ardenne and the city of Reims. The combination of farming co-operatives, private industry and backing through regional and national public policy and funding is perhaps the optimal model that can be reproduced in many locations.

Further added value has been created through an industrial ecology network. The end-of-pipe philosophy is clearly insufficient to prevent pollution. Equally, cleaner production has its limits. The industrial ecology approach considers, in the absence of a viable cleaner production alternative, using waste as a marketable by-product. Using waste from one process as an input to another process at the same site removes transportation and waste disposal or treatment costs.

Figure 5.7. **Business units at Bazancourt-Pomacle**

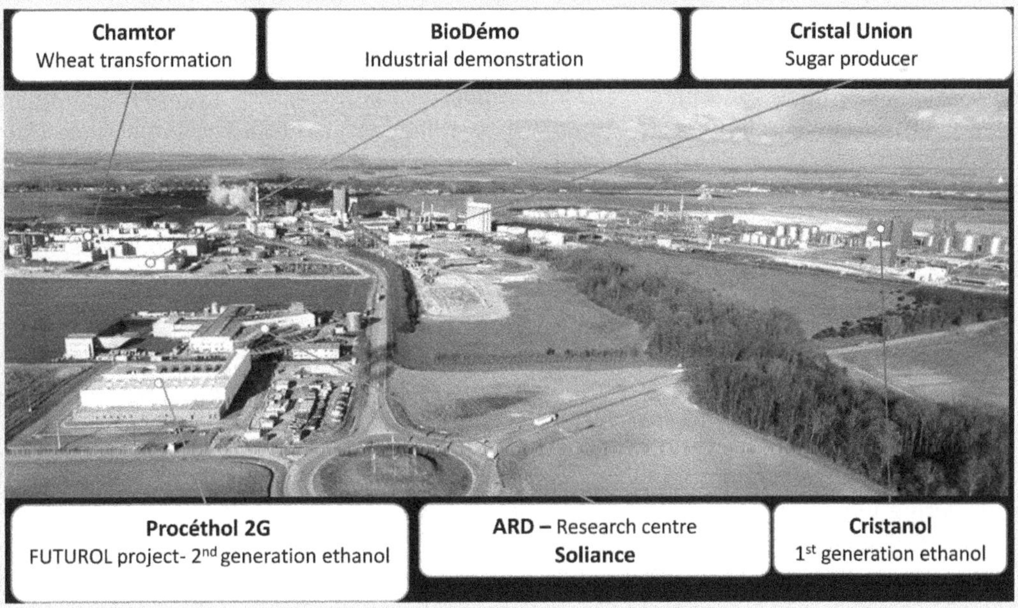

Source: Schieb and Philp (2014), "Biorefinery policy needs to come of age".

The most advanced wood biorefineries are found in Scandinavian countries. Borregaard (Norway), for example, boasts the most advanced biorefinery in the world. It has been making vanillin – one of the most valuable products made from wood – for more than 50 years (Borregaard, n.d.). Each year, it produces 1 500 tonnes from spruce wood.

In 2009, Chempolis, a Finland-based biorefining technology corporation, commissioned a biorefinery in Finland. Operating as a technology centre for testing customer raw materials for bioethanol, biochemicals and paper-making fibres, it has been also called the world's first demonstrator "third generation" wood-to-ethanol biorefinery.

In the northern portion of the Russian Federation, the Komi Republic could host a plant that would produce 100 000 tonnes of bioethanol per year from wood waste (Il Bioeconomista, 2016). The total investment required is estimated at EUR 136 million. A process to create a pool of investors is underway with different options under consideration, including a public-private partnership. Under the plans, the facility would process up to 400 000 tonnes per year of feedstock such as unusable timber and sawmill residues. The Komi Republic has rich forest resources, and local authorities have proposed a site of 15.6 acres for the plant.

Wastewater biorefineries

Probably the most widespread application of biotechnology worldwide is biological wastewater treatment technology. The core technologies have an unparalleled role in pollution prevention. Yet, in developing countries, 90% of sewage and 70% of industrial wastes are discharged without treatment into surface water. Wastewater management would play a central role in achieving future water security in a world with increased water stress (UN-Water, 2015).

With over a century of experience in biological wastewater treatment, advances beyond basic biochemical oxygen demand (BOD) and chemical oxygen demand (COD) removal are available. Small, modular systems requiring minimal civil engineering and maintenance are ideal for small, remote communities, while highly intensive plants can cater to city-sized populations. It would appear that large problems could be solved simply with greater implementation of biological wastewater treatment technologies (El-Chichakli et al., 2016). However, two points are worth bearing in mind.

1. Converting biodegradable materials in wastewater into non-toxic biomass, water and CO_2 has no added value.
2. Treatment of municipal wastewater accounts for approximately 3% of global electricity consumption and 5% of non-CO_2 GHG emissions, principally methane from anaerobic digestion (Li et al., 2015). In many cases, large wastewater treatment plants are the largest energy-consuming facilities in a city.

Future wastewater biorefinery models could well be derived from promising R&D. Considering the energy content embedded in wastewater is two to four times the energy used for treatment, future utilities could become energy-positive with the development of energy recovery technologies (McCarty et al., 2011). Moreover, these facilities could also recover other value-added resources such as nutrients, metals, chemicals and clean water. In this way, they could become closed loop waste biorefineries of very high productivity and efficiency (Lu and Ren, 2016), and potentially carbon-negative. Although global stocks of phosphate for fertilisers are being depleted, nutrients such as the phosphates and nitrogenous pollutants in wastewater contribute to disastrous instances of eutrophication.

Plastics from wastewater

Research is demonstrating how the organic carbon present in domestic wastewater can be converted by mixed microbial cultures into polyhydroxyalkanoate (PHA) bio-based plastics. These plastics are biodegradable with a range of functions that can replace traditional fossil-based plastics. Over 22 months, the Brussels North Wastewater Treatment Plant operated a pilot-scale biorefinery process to evaluate PHA production integration with services of municipal wastewater and sludge management (Morgan-Sagastume et al., 2015). Full-scale demonstration of the complete value chain alongside continuous polymer production remains to be validated (Paillard, 2016). Currently, this technology is at TRL 6.

Microbial electrolysis cells: Electricity from wastewater

Microbial electrolysis cells (MECs) can theoretically convert any biodegradable waste into H_2, biofuels and other value-added products. Since their invention in 2005 (Kadier et al., 2016), research has increased the H_2 production rate and yield by several orders of magnitude. However, many challenges must be overcome for MECs to be applied in large-scale systems (Randolph and Studer, 2013).

MEC technology can, in theory, be integrated into lignocellulosic biorefining. These biorefineries produce large amounts of wastewater that contain biodegradable organics. These can be used in MECs for additional energy production (Zeng et al., 2015).

Hungarian researchers (Szollosi et al., 2016) have developed a microbial fuel cell technology. It can produce a low-alcohol beer while it generates small amounts of electricity. Perhaps one day it will be possible to brew beer from wastewater in an energy-positive and carbon-negative process.

In Canada, Metro Vancouver (23 local authorities in the province of British Columbia) is working with Genifuel to build a demonstration plant that can convert raw sewage into biocrude oil. The technology is being licensed from the Department of Energy's Pacific Northwest National Laboratory (Ramirez, 2016).

Biogas biorefining

Anaerobic digestion of sewage sludge to produce biogas – a mixture of hydrogen, methane and carbon dioxide – has been used for over a century in the biological treatment of wastewater. Typically, it stabilises sewage sludge by removing pathogens. However, methane is typically used to generate electricity. This can often be enough to power an entire wastewater treatment plant, adding to the environmental and economic sustainability of such plants.

Anaerobic digestion is highly scalable. It has been perfected down to individual farm level, where a variety of waste materials can be converted to biogas (e.g. sludge, grass, solid manure, chicken manure and straw). Moreover, the effluents after anaerobic digestion are better balanced to meet crop needs than raw manure slurries. This reduces the need for supplementary chemical nitrogen and phosphorus fertilisers (Massé et al., 2011), while reducing GHG emissions (Siegmeier et al., 2015).

Biogas production is seen as part of the biorefinery concept (Kaparaju et al., 2009). Multiple biofuels production from, say, wheat straw (bioethanol, biohydrogen and biogas) can increase the efficiency of biomass use enshrined within the cascading use of the biomass concept. Volatile fatty acids (VFAs) produced from anaerobic microbial activity are often considered a nuisance or environmentally damaging. Yet they have potential as precursors for the biotechnological production of PHAs as bio-based plastics (Martinez et al., 2016).

The Centre for Advanced Sustainable Energy in Northern Ireland funds the BioGas to BioRefinery research project (QUB, n.d.). It aims to produce an evidence-based roadmap to develop a bioeconomy there. On the one hand, the research reviews the potential of feedstocks for biogas production in the country. On the other, it demonstrates the environmental and economic benefits of advanced use options for biogas from wastes.

Food waste biorefining

Roughly one-third of food produced for human consumption is lost or wasted globally, which amounts to about 1.3 billion tonnes per year (FAO, 2011). Further, the energy used in producing the food is also wasted. This means the GHG emissions associated with the production have been released with no benefit at all. Food that is produced but not eaten adds 3.3 billion tonnes of GHGs to the planet's atmosphere This makes food wastage the third top emitter after the total emissions of the United States and China (FAO, 2013).

TerraServ, a South African start-up formed in 2014, is developing a process to biologically convert food wastes into products such as hand sanitisers, whiteboard cleaners and glass cleaners under the brand name EcoEth (TerraServe, n.d.). The feedstocks are generally off-specification foods from manufacturers, goods damaged in transit or past their sell-by date. In the current phase of development (mid-2016), the company processes around 200 kg of food waste per month. Within the next year, it intends to increase this to 1-12 tonnes per month. The process is based on fermentation to ethanol. Ultimately, it aims to recycle wastewater and employ biological wastewater treatment, and to use as much solar heating as possible to minimise the carbon footprint (Coetzee, 2016).

Enterra of British Columbia, Canada, takes food waste from farmers, grocery stores and food producers in Metro Vancouver and the Fraser Valley, and feeds it to voracious black soldier fly larvae (Enterra, n.d.). In turn, the larvae can be processed into fertiliser and animal feed ingredients. Canada has recently approved this approach for chicken feed.

Note

1. Organic chemistry can be defined as a chemistry sub-discipline involving the scientific study of the structure, properties and reactions of organic compounds and organic materials, i.e. matter in its various forms that contain carbon atoms. However, biorefining most normally refers to "renewable" carbon.

References

Accenture (2009), *Betting on Science: Disruptive Technologies in Transport Fuels*, Accenture, New York.

Biodiesel Magazine (2016), "Ensyn biocrude receives CARB approvals for refinery coprocessing", 16 February, www.biodieselmagazine.com/articles/776356/ensyn-biocrude-receives-carb-approvals-for-refinery-coprocessing (accessed 17 January 2018).

Borregaard (n.d.), "The flavor that carries – Vanillin for 50 Years", webpage, Borregaard, Sarpsborg, Norway, www.borregaard.com/News/The-flavor-that-carries-Vanillin-for-50-years (accessed 7 December 2017).

Bioökonomierat (2016), "Wood in the bioeconomy – Opportunities and limits", *BÖRMEMO*, No. 5, Bioökonomierat, Berlin.

Bozell, J.J. (2008), "Feedstocks for the future – Biorefinery production of chemicals from renewable carbon", *CLEAN- Soil, Air, Water*, Vol. 368, Wiley Online Library, pp. 641-647.

BRBIOTECH-CEBRAP (2011), *Brazil Biotech Map 2011*, Brazilian Association of Biotechnology and Brazilian Centre for Analysis and Planning, Brasilia, www.clustercollaboration.eu/sites/default/files/international_cooperation/brazil_biotec_map_2011.pdf.

Cardona, C.A. et al. (2010), "Production of bioethanol from sugar cane bagasse: Status and perspectives", *Bioresource Technology*, Vol. 101/13, Elsevier, Amsterdam, pp. 4754-4766.

Carriquiry, M.A. et al. (2011), "Second generation biofuels: Economics and policies", *Energy Policy*, Vol. 39/7, Elsevier, Amsterdam, pp. 4222-4234.

Cheali, P. et al. (2015), "Upgrading of lignocellulosic biorefinery to value added chemicals: Sustainability and economics of bioethanol-derivatives", *Biomass & Bioenergy*, Vol. 75, Elsevier, Amsterdam, pp. 282-300.

Clark, P. (2015), "Microbes from rabbit gut help turn waste to fuel", *Financial Times*, 12 July, https://www.ft.com/content/788d87ec-273b-11e5-bd83-71cb60e8f08c.

Coetzee, W. (2016), "Sweet dreams", *The Chemical Engineer*, Vol. 899, Institution of Chemical Engineers, London, pp. 30-33.

de Jong, E. and G. Jungmeier (2015), "Biorefinery concepts in comparison to petrochemical refineries", in *Industrial Biorefineries and White Biotechnology*, Elsevier, Amsterdam, http://dx.doi.org/10.1016/B978-0-444-63453-5.00001-X.

Dias, M.O.S. et al. (2013), "Biorefineries for the production of first and second generation ethanol and electricity from sugar cane", *Applied Energy*, Vol. 109/C, Elsevier, Amsterdam, pp. 72-78.

Ducat, D.C. et al. (2011), "Engineering cyanobacteria to generate high-value products", *Trends in Biotechnology*, Vol. 29/2, Elsevier, Amsterdam, pp. 95-103.

El-Chichakli, B. et al. (12 July 2016), "Five cornerstones of a global bioeconomy", Nature Comment, Vol. 535/7611, Nature Publishing Group, London, pp. 221-223.

Emami, K. et al. (2015), "Proteomic-based biotyping reveals hidden diversity within a microalgae culture collection: An example using *Dunaliella*", *Scientific Reports*, Vol. 5/10036, Nature Publishing Group, London.

Enterra (n.d.), Enterra website, www.enterrafeed.com (accessed 7 December 2017).

FAO (2013), *Food Wastage Footprint: Impacts on Natural Resources*, Food and Agriculture Organization of the United Nations, Rome, www.fao.org/docrep/018/i3347e/i3347e.pdf.

FAO (2011), *Global Food Losses and Food Waste – Extent, Causes and Prevention*, Food and Agriculture Organization of the United Nations, Rome.

Federal Government of Germany (2012), "Biorefineries roadmap", Federal Ministry of Food, Agriculture and Consumer Protection, Berlin, May.

Giuliano, A. et al. (2016), "Process optimization of a multi-product biorefinery: The effect of biomass seasonality", *Chemical Engineering Research and Design*, Vol. 107, Elsevier, Amsterdam, pp. 236-252.

IEA Bioenergy Task 42 Biorefinery (2012), "Bio-based chemicals. Value added products from biorefineries", International Energy Agency, Paris, www.ieabioenergy.com/publications/bio-based-chemicals-value-added-products-from-biorefineries.

IEA Bioenergy Task 39 (2011), "Algae as a feedstock for biofuels. An assessment of the current status and potential for algal biofuels production", International Energy Agency, Paris, www.ieabioenergy.com/publications/algae-as-a-feedstock-for-biofuels.

Il Bioeconomista (6 February 2016), "In Northern Russia a plant to produce 100 000 tonnes of bioethanol per year from wood waste", Il Bioeconomista blog, https://ilbioeconomista.com/2016/02/06/in-northern-russia-a-plant-to-produce-100000-tonnes-of-bioethanol-per-year-from-wood-waste.

INEOS (31 July 2013), "First facility in the world using new advanced bioenergy technology to convert waste to renewable fuel and electricity," INEOS News blog, www.ineos.com/news/ineos-group/ineos-bio-produces-cellulosic-ethanol-at-commercial-scale.

Jungmeier, G. (2014), *The Biorefinery Complexity Index*, IEA Bioenergy – Task 42 Biorefinery, International Energy Agency, Paris, https://www.iea-bioenergy.task42-biorefineries.com/upload_mm/6/2/f/ac61fa53-a1c0-4cbc-96f6-c9d19d668a14_BCI%20working%20document%2020140709.pdf.

Kadier, A, et al. (2016), "A comprehensive review of microbial electrolysis cells (MEC) reactor designs and configurations for sustainable hydrogen gas production", *Alexandria Engineering Journal*, Vol. 55/1, pp. 427-443.

Kamm, B. et al. (2007), "Biorefineries – industrial processes and products", *Ullmann's Encyclopedia of Industrial Chemistry, 7th ed*, Wiley-VCH, Weinheim, Germany.

Kamm, B. et al. (2006), "Biorefinery systems – an overview", in Kamm, B. et al (eds.), *Biorefineries – Industrial Processes and Products: Status Quo and Future Directions, Vol. 1*, Wiley-VCH, Weinheim, Germany.

Kamm, B. et al. (eds.) (1998), "Die grüne Bioraffinerie" [The Green Biorefinery], in *Technologiekonzept*, Proceedings of 1st international symposium, October 1997, Berlin.

Kaparaju, P. et al. (2009), "Bioethanol, biohydrogen and biogas production from wheat straw in a biorefinery concept", *Bioresource Technology*, Vol. 100/9, Elsevier, Amsterdam, pp. 2562-2568.

Klein-Marcuschamer, D. et al. (2013), "A matter of detail: Assessing the true potential of microalgal biofuels", *Biotechnology and Bioengineering*, Vol. 110/9, Wiley Online Library, pp. 2317-2322.

Köpke, M. et al. (2011), "2,3-Butanediol production by acetogenic bacteria, an alternative route to chemical synthesis, using industrial waste gas", *Applied and Environmental Microbiology*, Vol. 77/15, American Society for Microbiology, Washington, DC, pp. 5467-5475.

Latif, H. et al. (2014), "Trash to treasure: Production of biofuels and commodity chemicals via syngas fermenting microorganisms", *Current Opinion in Biotechnology*, Vol. 27, Elsevier, Amsterdam, pp. 79-87.

Li, W.-W. et al. (2 December 2015), "Chemistry: Reuse water pollutants", *Nature* News and Comment, Vol. 528/7580, Nature Publishing Group, London, pp. 29-31.

Lu, L. and Z.J. Ren (2016), "Microbial electrolysis cells for waste biorefinery: A state of the art review", *Bioresource Technology*, Vol. 215, Elsevier, Amsterdam, pp. 254-264.

Lynd, L.E. et al. (2005), "Strategic biorefinery analysis: Analysis of biorefineries", *Report*, National Renewable Energy Laboratory, NREL/SR-510-35578, Golden, United States.

Martinez, G.A. et al. (2016), "Towards multi-purpose biorefinery platforms for the valorisation of red grape pomace: Production of polyphenols, volatile fatty acids, polyhydroxyalkanoates and biogas", *Green Chemistry*, Vol. 18/1, Royal Society of Chemistry, Cambridge, pp. 261-270.

Massé, D.I. et al. (2011), "On farm biogas production: A method to reduce GHG emissions and develop more sustainable livestock operations", *Animal Feed Science and Technology*, Vols. 166-167, Elsevier, Amsterdam, pp. 436-445.

McCarty, P.L. et al. (2011), "Domestic wastewater treatment as a net energy producer – can this be achieved?", *Environmental Science and Technology*, Vol. 45/17, American Chemical Society, Washington, DC, pp. 7100-7106.

Morgan-Sagastume, F. et al. (2015), "Integrated production of polyhydroxyalkanoates (PHAs) with municipal wastewater and sludge treatment at pilot scale", *Bioresource Technology*, Vol. 181, Elsevier, Amsterdam, pp. 78-89.

OECD (2017), *The Next Production Revolution*, OECD Publishing, Paris, http://dx.doi.org/10.1787/9789264271036-en.

OECD (2015), *Greening Steel: Innovation for Climate Change Mitigation in the Steel Sector*, OECD Publishing, Paris, https://www.oecd.org/sti/ind/Environmental-patents-steel.pdf.

OECD (2014), "Biobased chemicals and plastics: Finding the right policy balance", *OECD Science, Technology and Industry Policy Papers*, No. 17, OECD Publishing, Paris, http://dx.doi.org/10.1787/5jxwwfjx0djf-en.

Paillard, H. (2016), "Ever decreasing circles", *The Chemical Engineer*, Vol. 899, Institute of Chemical Engineers, London, pp. 38-40.

Parajuli, R. et al. (2015), "Biorefining in the prevailing energy and materials crisis: A review of sustainable pathways for biorefinery value chains and sustainability assessment methodologies", *Renewable and Sustainable Energy Reviews*, Vol. 43, Elsevier, Amsterdam, pp. 244-263.

Pavanan, K.C. et al. (2013), "Biomass sustainability and certification", *Trends in Biotechnology*, Vol. 31/7, Cell Press, Cambridge, United States, pp. 385-387.

Peters, D. (2011), "The German biorefinery roadmap", presentation at the expert forum conference on biorefineries", Budapest, April.

Pienkos, P.T. (2009), "Algal biofuels: Ponds and promises", presentation at the 13th annual symposium on industrial and fermentation microbiology, Lacrosse, United States, 1 May.

Pienkos, P.T. and A. Darzins (2009), "The promise and challenges of microalgal-derived biofuels", *Biofuels, Bioproducts and Biorefining*, Vol. 3/4, Wiley Online Library, pp. 431-440.

QUB (n.d.), "Biogas to Biorefinery," Queen's University Belfast, webpage, https://qubbiorefinery.wordpress.com (accessed 7 December 2017).

Ramirez, V.B. (18 December 2016), "Canada's building a plant that will make fuel out of poop", Singularity Hub In Focus blog, https://singularityhub.com/2016/12/18/canadas-building-a-plant-that-will-make-fuel-out-of-poop/#sm.0001nm9xhr5mkemergz1gqcgx0yr9.

Randolph, K. and S. Studer (2013), "Biological hydrogen production workshop summary report", US Department of Energy, Office of Energy Efficiency and Renewable Energy, Fuel Cell Technologies Office, Washington, DC.

Sayre, R. et al. (2013), "Initial risk assessment of genetically modified (GM) algae for commodity-scale cultivation", *Algal Research*, Vol. 2/1, Elsevier, Amsterdam, pp. 66-77.

Schieb, P.-A. and J.C. Philp (2014), "Biorefinery policy needs to come of age", *Trends in Biotechnology*, Vol. 32, Elsevier, Amsterdam, pp. 496-500.

Schieb, P.-A. et al. (2015), *Biorefinery 2030: Future Prospects for the Bioeconomy*, Springer Verlag, Berlin, Heidelberg, http://dx.doi.org/10.1007/978-3-662-47374-0.

Siegmeier, T. et al. (2015), "Farm biogas production in organic agriculture: System implications", *Agricultural Systems*, Vol. 139/C, Elsevier, Amsterdam, pp. 196-209.

Sjostrom, E. (1993), *Wood Chemistry (Second Edition): Fundamentals and Applications*, Elsevier, Amsterdam.

Soybean and Corn Advisor (2015), "Number of sugar/ethanol mills in Brazil continues to decline", 19 March 2015, *Soybean and Corn Advisor*, Hinsdale, United States, www.soybeansandcorn.com/news/Mar19_15-Number-of-Sugar-Ethanol-Mills-in-Brazil-Continues-to-Decline.

Star-COLIBRI (2011), *Joint European Biorefinery Vision for 2030: Strategic Targets for 2020 – Collaboration Initiative on Biorefineries*, Star-COLIBRI, Brussels.

Starkey, J. (2008), "High soy prices idling biodiesel plants", 13 May, *World Press Review*, www.public.iastate.edu/~nscentral/mr/08/0516/highsoy.html.

Subba Rao, P.V. and V.A. Mantri (2006), "Indian seaweed resources and sustainable utilization: Scenario at the dawn of a new century", *Current Science*, Vol. 91/2, Current Science Association, Bengaluru, India, pp. 164-174.

Sudhakar, M.P. et al. (2016), "Characterization, pretreatment and saccharification of spent seaweed biomass for bioethanol production using baker's yeast", *Biomass and Bioenergy*, Vol. 90, Elsevier, Amsterdam, pp. 148-154.

Szollosi, A. et al. (2016), "Production of low or non-alcoholic beer in microbial fuel cell", *Food and Bioproducts Processing*, Vol. 98, Elsevier, Amsterdam, pp. 196-200.

Tan, T., Yu, J. and Shang, F. (2011), "Biorefinery engineering" in *Comprehensive Biotechnology (Second Edition)*, Moo-Young, M. (ed.), Elsevier, Amsterdam.

Tanaka, K. et al. (1995), "Production of poly(D-3-hydroxybutyrate) from CO_2, H_2 and O_2 by high cell density autotrophic cultivation of *Alcaligenes eutrophus", Biotechnology and Bioengineering*, Vol. 45, pp. 268-275.

TerraServe (n.d.), TerraServe website, www.ecoeth.co.za/Our-Products/EcoEth (accessed 7 December 2017).

Tsakalova, M. et al. (2015), "A decision support environment for the high-throughput model-based screening and integration of biomass processing paths", *Industrial Crops and Products*, Vol. 75, Elsevier, Amsterdam, pp. 103-113.

UK Bioenergy Strategy (2012), Department of Energy & Climate Change website, www.gov.uk/government/organisations/department-of-energy-climate-change (accessed 7 December 2017).

UN-Water (2015), "Wastewater management", *Analytical Brief*, UN-Water, Geneva.

Wei, N. et al. (2013), "Marine macroalgae: An untapped resource for producing fuels and chemicals", *Trends in Biotechnology*, Vol. 31/2, Elsevier, Amsterdam, pp. 70-77.

Wikipedia (n.d.), "The Nelson Complexity Index", webpage, https://en.wikipedia.org/wiki/Nelson_complexity_index (accessed 7 December 2017).

Wright, M. and R. Brown (2007), "Comparative economics of biorefineries based on the biochemical and thermochemical platforms", *Biofuels, Bioproducts, and Biorefining*, Vol. 1/1, Wiley Online Library, pp. 49-56.

Zeng, X. et al. (2015), "Biotransformation of furanic and phenolic compounds with hydrogen gas production in a microbial electrolysis cell", *Environmental Science and Technology*, Vol. 49/22, American Chemical Society, Washington, DC, pp. 13667-13675.

Chapter 6.

Financing biorefineries

This chapter concentrates on the "commercialisation" and "scalable production" phases of biorefineries, i.e. demonstration and full-scale production. First-of-kind projects are high risk, and financing options dry up as a project approaches "scalable production". The early stage in the era of biorefining faces multiple barriers, including technical issues, public opinion, lack of supply, and value chains and lack of trained personnel. Small- to medium-scale biorefining is often cast as a rural manufacturing activity to be close to feedstocks such as agricultural products and residues and forestry. However, this smaller-scale distributed model competes directly with some of the largest global (and fossil-based) manufacturing industries. These factors, alongside lack of confidence in public policy, add up to high levels of financial risk in biorefineries for the private sector. This has led to various forms of public-private partnerships to give the private sector concrete commitment from governments in order to make biorefineries a production mode of the future.

Introduction

In 2017, certain categories of biorefineries are still high-risk investments. There are at least hundreds, perhaps upwards of 1 000, first-generation bioethanol mills worldwide. In Brazil, these function according to the market, but in competition with gasoline. Thus, the light vehicle fleet is overwhelmingly flex fuel, allowing the consumer to choose between ethanol and gasoline according to the local spot price.

The greatest financial risk, however, is with biorefineries favoured most by policy makers. These are second-generation, cellulosic biorefineries that primarily, at this stage, produce second-generation ethanol (Peplow, 2014). Policy makers favour these biorefineries because they appear to tackle some of the toughest of the so-called grand challenges: climate change, energy security and resource depletion. However, cellulosic biorefineries tackle an additional policy – the need to avoid competition for land. Since feedstocks are intended to be residues and waste materials rather than food or feed crops, cellulosic biorefineries do not create a competition between food and industrial needs for land. Late in April 2015, the European Parliament passed a draft law to cap crop-derived biofuel production and accelerate the shift to alternative sources (Il Bioeconomista, 2015).

In the United States, the public agencies most directly involved in biorefining are the Department of Agriculture (USDA) and the Department of Energy (US DOE). Both agencies help finance biorefineries by the same instrument – the loan guarantee. In Europe, the situation is radically different, and a little more background is required to explain this situation.

A gap in the innovation cycle has long existed in Europe, which has fewer top class research establishments and less infrastructure and capability (OECD, 2015). It has also been seen to be much less capable at commercialising the results of promising research. The Bayh-Dole Act of 1980 set public research establishments in the United States on a clear path to commercialisation of research and development (R&D) (McManis and Noh, 2006). It permitted a university, small business or non-profit institution to pursue ownership of an invention rather than to keep it in the hands of government.

Funding mechanisms for industrial biotechnology and biorefinery projects are summarised in Figure 6.1.

Figure 6.1. **Funding mechanisms in industrial biotechnology**

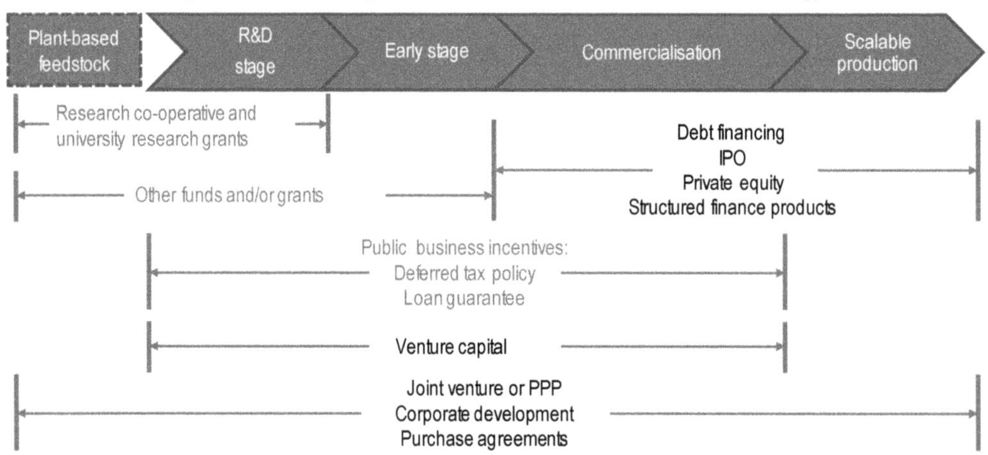

Notes: PPP = public-private partnership; blue text = public sources; IPO = initial public offering.

Source: Adapted from Milken Institute (2013), *Unleashing the Power of the Bio-economy.*

Demonstrator plants can help prevent mistakes in full-scale production

Demonstrator plants are larger than pilot plants, but smaller than full-scale production plants (Figure 6.2). Many of the technical, supply chain and economic issues become apparent, and can therefore be addressed, at the demonstrator phase. It is therefore a vital stage to prevent expensive mistakes from occurring at the full-scale production phase. This option is also suitable for gaining experience from small-scale experiments aimed at attracting potential participants' interests (e.g. investors, credit institutes, local public policy makers, suppliers of raw materials, final users) who are not yet aware of the opportunities deriving from the new business. Yet demonstrator plants are notoriously difficult to finance, a barrier that could be addressed through public-private partnerships (PPPs) or other public support mechanisms.

Figure 6.2. **Three scales of bio-based production**

A. Pilot

B. Demonstration

C. Full scale

Sources: (a) Courtesy of Pierre Guerin Technologies, France; (b) courtesy of the Centre for Process Innovation, United Kingdom; (c) Peplow, M., "Cellulosic ethanol fights for life: Pioneering biofuel producers hope that US government largesse will ease their way into a tough market".

Whether demonstrator or full commercial scale, the costs involved are much larger than those allocated to research programmes, and thus require private-sector involvement. The current answer in Europe is the European bio-based industries PPP. After years of planning and development, the PPP was finally launched on 27 June 2014 with a budget of EUR 3.7 billion for bio-based innovation from 2014-20 (Horizon 2020). Of the total, EUR 975 million is EU funds, with the other EUR 2.7 billion from private investments.

Coetzee (2016) listed the advantages TerraServ (South Africa) considers from piloting its process, which apply equally to demonstrators. First, it has helped identify pitfalls in the process, as well as critical needs for control and instrumentation (as these can make up a significant amount of the total capital required). It has helped optimise the process from technical and business perspectives. In addition, it has been useful to develop novel technology needed in a full-scale process. It is also a tangible asset that can be shown to stakeholders and potential investors. Finally, it produces an actual product that can help establish market interest before committing capital to a full-scale process.

Figure 6.3 shows the funding of development-scale projects in four countries and the European Union, revealing a significant difference across the globe.

Figure 6.3. **International comparison of the share of basic, applied and development activities**

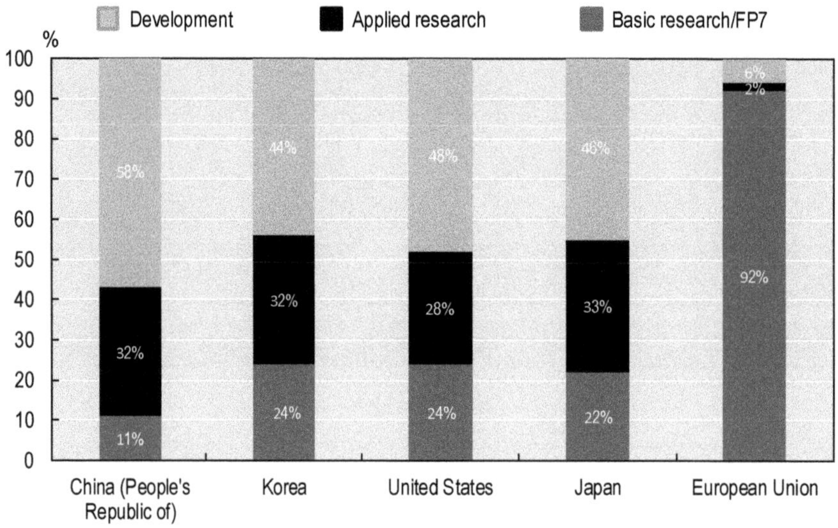

Note: FP7 = Framework Programme 7.

Source: Falholt, P., http://www.academia.edu/8097142/03_Per_Falholt.

Equity funding vs. traditional bank loans

The financing of such a complex construction project as a biorefinery has a dramatic effect on the project's execution. For example, the Crescentino cellulosic biorefinery had difficulties with debt management. Equity funding is the preferred approach from the engineering perspective and the quickest way to deliver the project. The company retains key decisions, like purchasing major equipment, avoiding the need for outside approval. This would be a preferred route for, say, a new petrochemicals plant. It would allow investors to know that the risks are low, and dividends are generally steady, if not spectacular, over long periods. However, for a first-of-kind biorefinery, especially for large expensive facilities, this approach is not common; the risks are too high, and it creates dilution for existing shareholders if significant capital is raised as equity.

At the other end of traditional project financing, a straight bank loan provides a large proportion of required funding. This normally carries significant restrictions and approvals. In the case of a biorefinery, there is significant risk in the feedstock supply chains; all contracts would require careful and time-consuming review. Also, all major equipment purchases would require sign-off. At least for first-of-kind biorefineries, not all of these details can be known at the time of fundraising. Even if this funding route were possible, it would require extended timelines for the many approvals; this, of course, increases costs. As more cellulosic biorefineries are built and have proven financial viability, this route may become more popular and realistic.

Clearly, then, different financing situations pertain to different stages of development of biorefineries. This is an important point for policy makers who must decide on the optimal public investment strategies for the different stages of the process (from basic R&D through to commercialisation). The critical differences are highlighted in Figure 6.4.

Figure 6.4. **Criteria determining bio-based investment decisions**

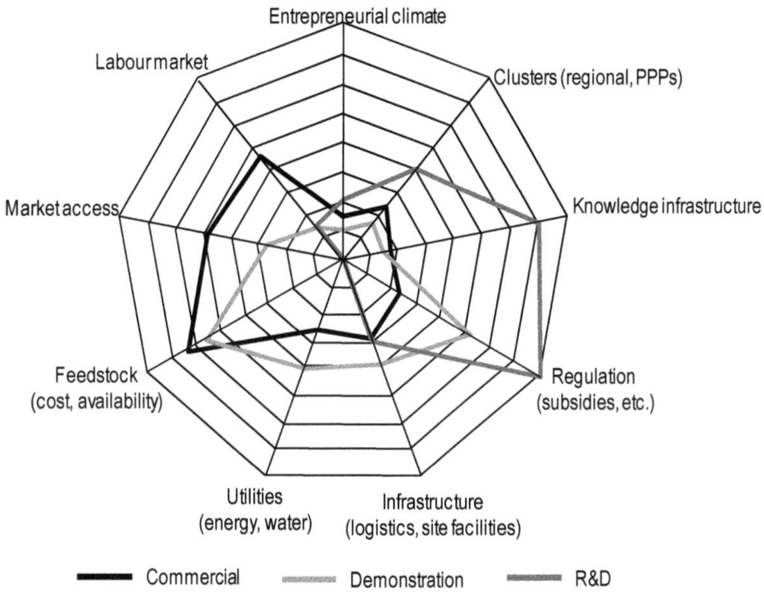

Note: PPP = public-private partnership; R&D = research and development.

Source: Suurs and Roelofs (2014), "Biobased investment climate in the Netherlands and Europe. Summary results quick scan", www.tno.nl/media/3387/quickscan_biobased_investment_climate_executive_summary.pdf.

In R&D, public research subsidy is, of course, the top priority, while feedstock issues may be irrelevant. However, feedstock costs are of critical importance at demonstration phase, and even more so at commercialisation. The other two criteria critical to determining commercial success are the labour market (ensuring an adequate supply of skilled labour) and market access (which can be influenced by public procurement).

Loan guarantees and debt finance

OECD (2003) recommends examining loan guarantees before loans: "The creativity, resources and know-how of markets can provide powerful support to policy. However, local and regional policy makers often overlook the role of markets."

The most common form of financing for such technologies in the United States is a hybrid of equity, teamed with either federal grants or federally backed loan guarantees. A grant does not need to be paid back, but is subject to a series of technical hurdles. To build biorefineries, both the USDA and US DOE have favoured 20-year loan guarantees.

With a government loan guarantee, the government (the guarantor) promises to assume the debt obligation of a private borrower if that borrower defaults. Loan guarantees are similar to traditional project finance, but the government accepts the technology risk and backs the loan. This streamlines the approval steps and the control. In the case of Crescentino, this may have removed many of the debt management problems.

New policy in the United States requires biorefineries to produce, but not sell, an advanced biofuel

Under new provisions of the assistance programme for biorefineries in the United States (Box 6.1), biorefineries must produce an advanced biofuel. However, they are not required to sell it as a biofuel. This removes a previous requirement that 51% of the product be sold as advanced biofuels. The options are noted below:

- It (the biorefinery) may sell the advanced biofuel that it produces as a biofuel, a renewable chemical or for other non-fuel usage.
- It may process the advanced biofuel into renewable chemicals or other bio-based products.
- It may use the biofuel as a fuel for heat or power in its processes or to generate electricity.

Box 6.1. **Changes to the USDA Farm Bill, Program 9003**

For the Farm Bill of 2014, Program 9003, the USDA Biorefinery Assistance Program was renamed the Biorefinery, Renewable Chemical, and Biobased Product Manufacturing Assistance Program. The USDA was directed to ensure diversity in the types of projects approved. It also had to cap the funds used for loan guarantees to promote bio-based product manufacturing at 15% of the total available mandatory funds. However, the same policy mechanism is now supporting both biofuels and bio-based products and materials. It provides loan guarantees up to USD 250 million.

Funds may be used to fund the development, construction and retrofitting of:

- commercial-scale biorefineries using eligible technology
- bio-based product manufacturing facilities that use technologically new commercial-scale processing and manufacturing equipment to convert renewable chemicals and other bio-based outputs of biorefineries into end-user products on a commercial scale
- refinancing, in certain circumstances.

Importantly, the programme distinguishes between biorefineries and bio-based manufacturing facilities. The terms and conditions, and eligibility are clearly set out (www.rd.usda.gov/programs-services/biorefinery-renewable-chemical-and-biobased-product-manufacturing-assistance). Federal participation (loan guarantee, plus other federal funding) cannot exceed 80% of total eligible project costs. The borrower and other principals involved in the project must make a significant cash equity contribution.

In broad terms, two types of projects are eligible within the programme: biorefineries and bio-based product manufacturing facilities (as these may be treated differently in the United States). This is a welcome development as it expands opportunities for new technologies, new processes and products. It also provides loan guarantees to bio-based product manufacturing facilities, whereas previously the programme was open only to biofuel biorefineries.

The application is now a two-phase process. Information required during phase one has been reduced to lessen the burden on applicants not selected for phase two. For scoring purposes, biorefineries and bio-based product manufacturing are ranked separately and compete against similar projects. Applications with the highest priority score rankings, in each category, are invited to submit a proposal for phase two.

InnovFin in Europe aims to improve access to risk finance

The InnovFin-EU Finance for Innovators (EIB, n.d.) was launched by the European Community and the European Investment Bank (EIB) Group in the framework of Horizon 2020. It provides guarantees or direct loans (EUR 24 billion available) to research and innovation projects. InnovFin aims to improve access to risk finance for research and innovation projects; research infrastructures; public-private partnerships; and special-purpose projects promoting first-of-a-kind, industrial demonstration projects (Scarlat et al., 2015). This is a major step in Europe as loan guarantees had previously been missing from the portfolio of funding mechanisms for bioeconomy projects. A summary of InnovFin products is shown on Table 6.1.

Table 6.1. **Overview of InnovFin products**

SMEs	Mid-caps	Large caps	Advisory
SME Guarantee[1]	MidCap Guarantee[1]	Large projects	
SME Venture capital	MidCap Growth finance		

1. Denotes indirect products. Direct products: the EIB group directly issues a loan to a borrowing project (loan covers up to 50% of total project costs). Indirect products: the EIB group offers (counter-)guarantees to an intermediary partner bank which then issues loans to borrowing projects ([counter-]guarantees cover up to 50% of project costs).

Note: There is no common EU definition of mid-cap companies. While small and medium-sized enterprises (SMEs) are defined as having fewer than 250 employees, mid-caps are broadly said to have between 250 and 3 000 employees.

Source: Fernández Gutiérrez (2016), "Bio-based industries in the European bioeconomy".

A bio-based industries instrument for Europe: BIC/BBI

The European bio-based industries PPP was launched in 2014. As noted previously, Europe is perceived as lagging behind in bioeconomy matters, especially when it comes to full-scale commercialisation. Good groundwork in basic research is then often capitalised upon abroad.

As one reason for the lag, the phases of the innovation chain beyond R&D are usually much more expensive. In the case of bio-based production, they are also very high risk. Governments lack policy stability and the drive towards establishing a bioeconomy. However, the private sector is unwilling to shoulder the entire financial burden in building the first key facilities without public support.

In another challenge for Europe, different industry sectors that should act together in bioeconomy development have been operating in isolation. These span agriculture, agri-food, forestry/pulp and paper, technology providers, chemicals and energy (Figure 6.5).

Figure 6.5. **The vision of the BBI for an integrated European bioeconomy**

(a) Where Europe is | (b) Where BBI wants it to be

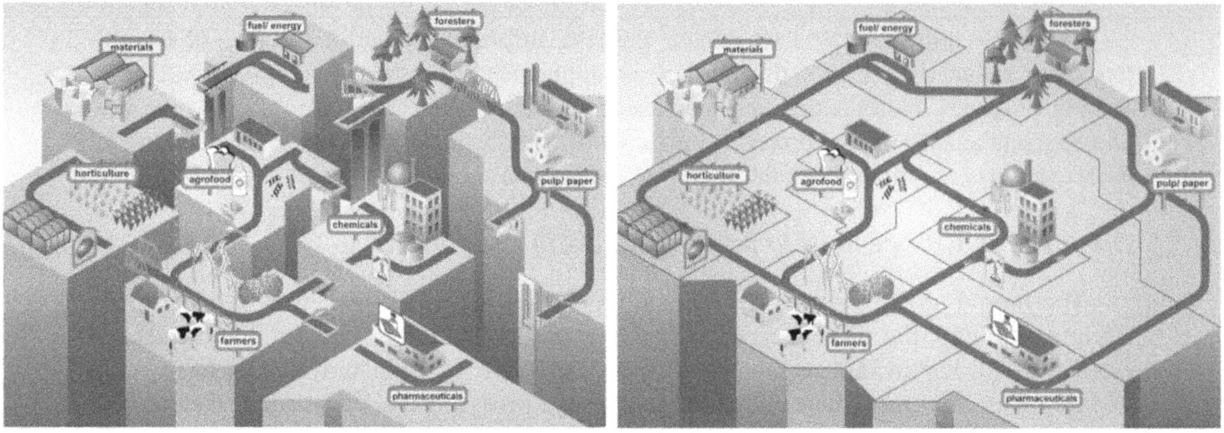

Source: Carrez (2014), "The bio-based industries initiative PPP and a focus on higher added-value".

The establishment of the Bio-based Industries Initiative (BBI) in 2014 is meant to address these issues (BBI, n.d.). The BBI supports industrial research and innovation to overcome the innovation "valley of death" – the path from research to the marketplace that has long been identified as a weakness in Europe. As of May 2016, membership consisted of 70 full members (large companies, SMEs and regional clusters) and 155 associate members (universities, regional trade organisations, European trade organisations, European Technology Platforms and private banks). The membership spans the critical sectors of a bioeconomy: agriculture, agri-food, forestry, pulp and paper, technology providers, chemicals, energy and others.

Having these diverse sectors represented is essential as the PPP focuses on the complete value chain. If a new value chain has gaps, it is impossible to operate properly. Figure 6.6 depicts the value chain concept in bio-production.

The core elements of the BBI strategy are: a robust framework that brings clarity for activities and investments; long-term stability and predictability; a joint approach, across sectors and across nations; joint financial commitment and a jointly defined programme that will unite parties that would otherwise find these activities to be too risky for an individual sector/company to carry out on its own; leveraging of further investments; industry driven and therefore result- and market-oriented.

A major task for BBI and its consortium is to level the fragmentation and build bridges between the sectors (Figure 6.5). A Strategic Innovation and Research Agenda (SIRA) that concentrates on value chains is expected to achieve this goal. SIRA suggests the following specific value chains (VCs) (BIC, 2013):

- VC 1: from lignocellulosic feedstock to advanced biofuels, bio-based chemicals and biomaterials, realising the feedstock and technology base for the next generation of fuels, chemicals and materials
- VC 2: next-generation forest-based value chains using the full potential of forestry biomass by improved mobilisation and realisation of new added value products and markets

- VC 3: next-generation agro-based value chains, realising the highest sustainability and added value by improved agricultural production, and new added value products and markets
- VC 4: new value chains from (organic) waste, from waste problems to economic opportunities by realising sustainable technologies to convert waste into valuable products
- VC 5: integrated energy, pulp and chemicals biorefineries realising sustainable bioenergy production, by backwards integration with biorefinery operations isolating higher added value components.

Sectors that have never collaborated before are creating new value chains. For example, the food industry is collaborating with the chemical industry (BBI, n.d.).

Figure 6.6. **Bio-based industrial value chains**

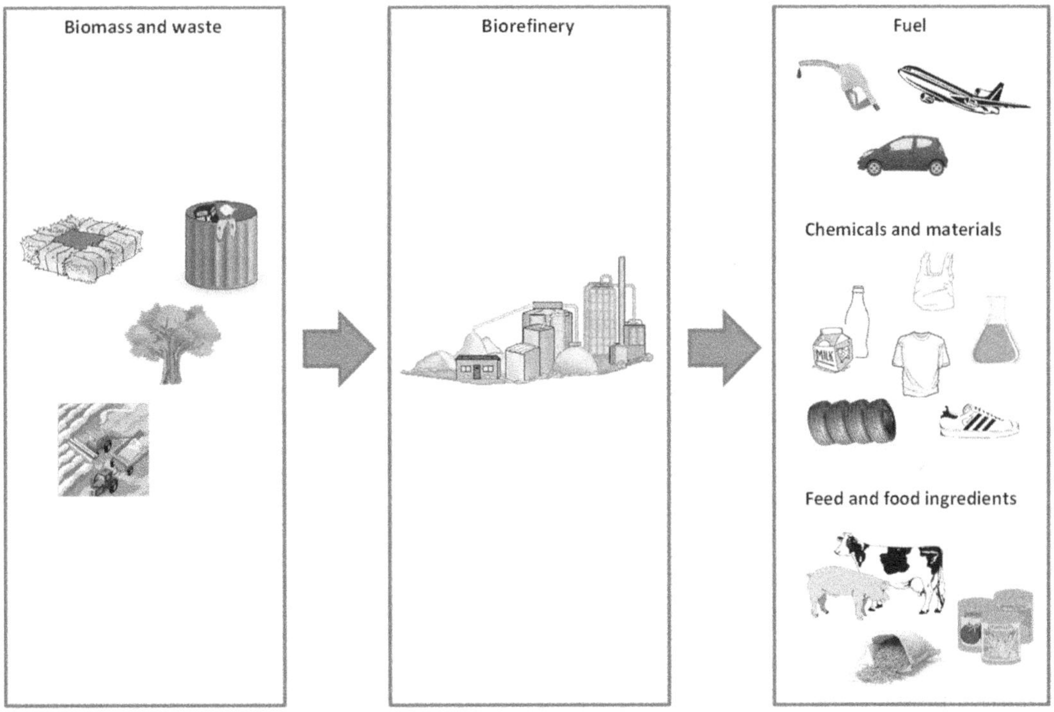

Source: Redrawn from Carrez (2016), "The current status of the BBI Joint Undertaking".

Focusing on higher added value

The consistent theme for SIRA is the emphasis on higher added value. This is how Europe wants to break away from concentrating on first-generation biofuels and bioenergy, which have consistently had lower added value and probably lower job creation potential. Through added value, SIRA can meet several of its biggest challenges – climate change obligations, energy security, rural regeneration, chemicals industry competitiveness, and overcoming unemployment and low growth.

Absolutely crucial to SIRA is its range of projects, which spans R&D, demonstration, flagship and supporting projects. To encourage more capacity building within Europe, flagship projects are meant to optimise technology for biomass conversion and ensure price competitiveness. They will both build new operations and upgrade existing and abandoned industrial sites by their conversion to biorefinery operations. Each value chain area will lead to at least one flagship project.

Investment in research and innovation, and skills

In the European Horizon 2020 programme, funding for bioeconomy research is available through Societal Challenge (SC) 2: Food Security, Sustainable Agriculture and Forestry, Marine and Maritime and Inland Water Research and the Bioeconomy. This is funded to the level of EUR 3.8 billion. Within its objectives is policy integration through increased multi-disciplinary and cross-sectoral research and innovation (R&I), smart specialisation (European Structural and Investment Funds, ESIF) and smart investment (the European Fund for Strategic Investments, EFSI). For example, EFSI mobilises funds in the bioeconomy arena through the EUR 75 million loan agreement between the European Investment Bank (EIB) and Metsäliitto Cooperative for the construction of a new large-scale bio-product mill in Finland.

Alternative instruments

Some more innovative instruments are emerging for what are seen to be "green" projects i.e. major environmental projects such as wind farms and solar energy. Their deployment in biorefinery projects to date is unknown, but is certainly not widespread yet. The Finland international biorefinery competition is directly aimed at bio-based production.

COP21 and Energy Union Integrated Research, Innovation and Competitiveness Strategy (EURICS)

The European Community recognises that deep decarbonisation of the economy is not possible without the bioeconomy. A new Research, Innovation and Competitiveness Strategy of the Energy Union (EURICS) was to be part of the 2016 State of the Energy Union, planned for November 2016 (EC, 2016). The strategy was to provide the Commission's follow-up to the outcome of COP21 as regards R&I.

The strategy explicitly identifies major innovations that are needed. Sustainably produced biomass must replace fossil resources for energy, fuels, chemicals and materials (plastics) – without compromising food security. Biomass production (agriculture) is a major source of GHG emissions (e.g. livestock, fertiliser production), but also a possible sink of CO_2 (e.g. forestry and soils).

Green bonds

Green bonds are a recent development with potential to become the major financing route for "green" projects. Green bonds enable capital-raising and investment for new and existing projects with environmental benefits (ICMA, n.d.). The Green Bond Principles instrument, for example, raises large capital sums. Project sponsors rather than investors take on financing and risk management.

Recently, a consortium of investment banks announced its support of the Green Bond Principles – Bank of America, Merrill Lynch, Citi, Crédit Agricole Corporate and Investment Banking, JPMorgan Chase, BNP Paribas, Daiwa, Deutsche Bank, Goldman Sachs, HSBC, Mizuho Securities, Morgan Stanley, Rabobank and SEB. The initiative is very new and still evolving. The Bank of America Corporation has announced that it has issued a "green bond" consisting of a three-year, fixed-rate bond that is USD 500 million in aggregate principal amount. This issuance of bonds is part of the company's ongoing commitment to advance renewable energy initiatives and promote energy efficiency.

The OECD now convenes the annual Green Investment Financing Forum (GIFF) (OECD, 2016a). The 3rd GIFF event was held in Tokyo in 2016, in association with the Asian Development Bank Institute. This event, as in previous years, gathered senior policy makers and key actors in financing green investment from around the world for a targeted discussion, this time themed on Asia. Participants discussed various issues, including mobilising private investment in low-carbon and climate-resilient infrastructure, and managing financial risks arising from climate change. They also explored challenges and opportunities for institutional investors, development of green bond markets, early stage equity finance and greening the traditional banking sector. Finally, they examined the role of public financial institutions, including public green banks; the potential for local and retail green finance; new and emerging actors in green finance; and policies and regulation to get on a low-emissions pathway.

UK Green Investment Bank

Over a dozen national (e.g. Australia, Japan, Malaysia, Switzerland, United Kingdom) and subnational governments have created public green investment banks (OECD, 2016b) primarily to reduce major barriers to scaling up low-carbon and climate-resilient infrastructure investment. In 2012, for example, the government of the United Kingdom created the UK Green Investment Bank plc. (UKGIB) to foster private-sector investment in projects related to environmental preservation and improvement. This followed the report of a non-partisan House of Commons committee on climate change that projected a funding gap of hundreds of billions of pounds (Sterling) for green infrastructure. It stated that traditional sources of capital for investment in green infrastructure (utility companies, project finance and infrastructure funds) could not provide even half the amount needed by 2025. As a result, the state budget would need to cover the balance (House of Commons, 2011). This led to the creation of UKGIB.

The bank differs from a typical "fund". Beyond disbursing government money, it raises its own finance and fills a gap in the market for government-backed bonds. This requires banking expertise and a range of commercially-driven interventions – loans, equity and risk reduction finance. To make such a mechanism viable, it must attract private sector investment and operate commercially without direct influence from the government. UKGIB, mandated to operate as a "for profit" bank, became operational in October 2012, with GBP 3 billion of UK taxpayer capital. An investment alliance with Abu Dhabi-backed clean energy firm Masdar aims to bring in additional funding to support UK projects over the next seven years.

UKGIB can invest in the following projects, many of which include bio-based production: large energy de-carbonisation projects; SMEs; innovation; new technologies and R&D; community-scale action; investment priorities; and nuclear power. In April 2017, the UK government agreed to the sale of the Green Investment Bank to the Australian bank Macquarie.

New York Green Bank Initiative

The USD 1 billion New York Green Bank Initiative opened in February 2014. Initially capitalised with USD 210 million in funding, the bank will partner with private-sector institutions to provide financing for qualifying clean energy projects and to accelerate clean energy deployment in the state of New York.

In its first request for proposal, the New York Green Bank invited clean energy industry participants to subject plans for various projects, including those related to energy generation and energy savings. These, in turn, may feature a wide range of commercially proven technologies, including biomass projects.

Bpifrance

La Banque publique d'investissement (Bpifrance) finances businesses of all sizes and stages – from the seed phase, to the transfer to stock exchange listing, and finally, to loans, guarantees and equity. Bpifrance accompanies firms developing export activities, in partnership with UBIFRANCE and Coface, and supports their innovation projects.

The French State and the Deposits and Consignment Fund (Caisse des Dépôts) are equal shareholders in Bpifrance. Therefore, it acts in support of public policy established by the state and the regions. In common with many French initiatives in industrial biotechnology, Bpifrance is highly supportive of regional innovation: 90% of its decisions are made regionally, where entrepreneurs are located.

SPIRE and Horizon 2020

The Sustainable Process Industry through Resource and Energy Efficiency (SPIRE) PPP is not dedicated to bio-based production, but its objectives fit with its policy goals and those of the bioeconomy more generally. SPIRE, in turn, is part of Horizon 2020, the EU framework programme for research and innovation. This framework, which runs from 2014 to 2020, comprises a EUR 80 billion budget.

SPIRE is a contractual PPP dedicated to innovation in resource and energy efficiency in the process industries (hence the fit to bio-based production). It develops the enabling technologies and solutions along the value chain to help Europe reach long-term goals of global competitiveness, and environmental and social sustainability. Eight industry sectors, including chemicals, have contributed to the development of SPIRE via European Technology Platforms and Industry Associations.

Box 6.2. **The PRODIAS project**

A consortium of companies in the European process industry from the areas of biotechnology, renewable resources, chemistry, process engineering, equipment supply and research organisations recently launched PROcessing Diluted Aqueous Systems (PRODIAS). The project focuses on unlocking the potential of renewable-based products made via industrial biotechnology. To that end, it envisions significantly decreasing production costs, increasing productivity and efficiency, lowering energy consumption and accelerating process developments.

Under the leadership of BASF, the partners include Cargill Haubourdin, France; University of Kaiserslautern, Germany; Imperial College London, United Kingdom; Alfa Laval, Sweden; GEA Messo PT and Exendo, the Netherlands; UPM, Finland and Enviplan, Germany. The goal is to develop cost- and energy-efficient technologies for water purification, removal and product-recovery needed to support downstream processing in industrial biotechnology.

The total project budget is about EUR 14 million with the European Union contributing EUR 10 million. EU funding of the PRODIAS project is enabled via the PPP with SPIRE (Sustainable Process Industry through Resource and Energy Efficiency).

The PRODIAS (PROcessing Diluted Aqueous Systems) (Box 6.2) project addresses downstream processing in bio-based production. In many bioprocesses, downstream processing can be extremely expensive, especially as it often produces large volumes of wastewater.

Biorefinery competitions

Competitions can simplify rules and regulations, and drive innovation. As part of its bioeconomy strategy, launched in 2014, Finland put in place an international biorefinery competition (Box 6.3). The competition sought to accelerate commercialisation of novel processes, as well as product and business innovations related to the bioeconomy, and to boost new biorefinery investments. Although modest in the cash investment involved, such an initiative may be important in leveraging other funding. As a public sector contribution, it should signal serious intent, and hopefully policy stability, to the private sector. Entrants to such a competition should also send signals to government about the types and diversity of activities involved. This information, for example, could be used in developing a national or regional biorefinery roadmap.

Box 6.3. Winners of the International Biorefinery Competition

In June 2014, Finland launched an international competition to expedite commercialisation of bioeconomy innovations and the emergence of new biorefineries. It is part of the drive by the Ministry of Employment and the Economy to implement the government's bioeconomy, cleantech and digitalisation strategy for accelerating new areas for growth. Three entrants to the International Biorefinery Competition won awards in 2015:

- Spinnova Ltd., for its new textile fibre production technology that makes it possible to spin yarn directly from wood fibre
- Biovakka Suomi, for its concept to combine production of biogas, nutrients and transport fuel
- The Kemijärvi Consortium, which incorporates novel Finnish technology to produce new biomaterials and biochemicals.

The prize money is relatively modest at EUR 100 000. However, a diverse array of proposals from different bioeconomy areas took part in the competition. They varied in size from demonstration plants worth less than EUR 1 million to biorefineries requiring hundreds of millions of euros in investment. The intent is that relatively small public investments leverage much larger private investments.

Sources: Adapted from VTT Ventures Ltd. (2015), "Spinnova developing environmentally friendly yarn thread technology based on spruce and pine fibres", www.vttresearch.com/media/news/spinnova-developing-environmentally-friendly-yarn-thread-technology-based-on-spruce-and-pine-fibres, and Finland Times (2015), "Spinnova's project wins bio-refinery competition", www.finlandtimes.fi/business/2015/02/26/14606/Spinnova%E2%80%99s-project-wins-bio-refinery-competition.

Where next for biorefinery finance?

The future for biorefinery finance would seem to be "more of the same". However, regarding public investments, specific areas would be worth tackling as priorities (Figure 6.7).

This analysis is consistent with the view that feedstock costs must be stabilised to encourage the private sector to build the necessary supply and value chains. As most of the jobs associated with biorefining are outside the plants, in the form of indirect and induced jobs, it also makes sense for policy makers to focus on this issue.

Capital cost is also a critical issue. The first of the flagship facilities is being built, but more need to follow, especially those using novel, non-food feedstocks. Long delays associated with cellulosic biorefineries attest to this need. However, an R&D subsidy is clearly still required in several key areas, especially conversion technologies.

The answer is also country-dependent. A national biorefinery roadmap, backed by a leadership council to make sure that milestones are met, would aid policy makers. Compiling such a roadmap requires countries to know what feedstocks are available locally and what needs to be imported (if anything).

Figure 6.7. **Key variables for policy support in biorefinery finance**

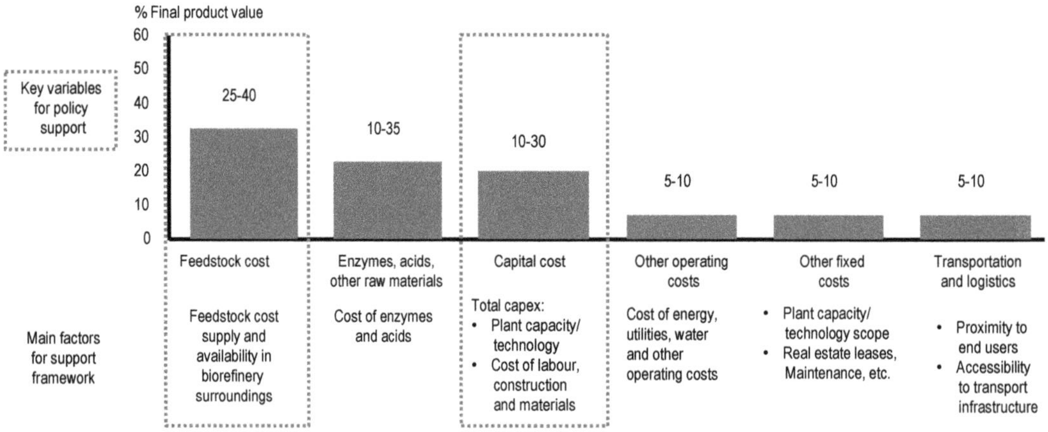

Note: capex = capital expenditure.

Source: Redrawn from Panoutsou (2015), "Integrated biorefineries and innovations in the optimal use of biomass".

Policy implications

For the private sector, the over-riding concern about bio-based production, independent of geography, relates to policy stability and uncertainty: governments need to make the explicit connection between uncertainty and investment risk. There are many worries; those generating the greatest uncertainty are outlined.

- Public perception: this has been influenced by years-long debates about indirect land-use change (ILUC) and "food vs. fuel". Fears about ILUC in Europe, at least, appear to be receding, but this may be temporary.
- Eligible feedstocks: there is apparent political support for cascading use of biomass as it aligns with the circular economy and zero waste. Therefore, the use of wastes and residues is supported, but available capacities for second-generation biofuels are uncertain.
- Availability and acceptability of energy crops for second-generation biofuels: the public policy concerns are typically about displacing food crops with non-food crops for industrial use, and the effects on food prices. This also then results in worries about available capacities for second-generation biofuels.
- Renewable Energy Directive (RED): looming behind the concerns already noted is what happens to RED after 2020.

- Lack of finance schemes for large projects: Officials from Biochemtex, owners of the Crescentino plant in Italy, have publicly stated that credit/loan guarantees (in addition to, or as an alternative to, grants) are needed.

Research centres and clusters should become more investment-savvy

Research centres and clusters must be more knowledgeable about investments. For example, they should know what types of financial instruments are available, nationally and regionally (even at the city level). They should identify venture capital firms of interest. They should understand the strengths of the research areas and individual researchers, as well as of local and wider small, medium and large companies with interests in industrial biotechnology (not necessarily dedicated biotechnology firms). Most crucially, they need to know how to partner different organisations and individuals. In this way, research centres and clusters can improve the chances of getting projects funded by taking away background work that others may be unwilling to do.

Box 6.4 profiles Toulouse White Biotechnology (TWB), an excellent example of where a consortium agreement that streamlines contracting and project management is the cornerstone of an investment strategy.

Box 6.4. **Financial instruments used at Toulouse White Biotechnology**

Toulouse White Biotechnology (TWB) is a pre-industrial demonstrator for sustainable production based on industrial biotechnology based in France. It is considered a "future centre of excellence in the field of industrial, or white, biotechnology". It aims to be a Joint Service Unit under the auspices of the National Institute for Agricultural Research, National Institute for Applied Sciences and National Centre for Scientific Research. There are 23 industrial partners and 9 public institutions involved, which adhere to a collaboration agreement that simplifies contract negotiations.

Collaborative academic-industry research programmes (pre-competitive) are co-financed by private companies, primarily through annual fees. Academic research organisations participate for free on projects.

Competitive projects are financed exclusively by private companies. Regional sources of funding are also important, but European funding can be involved as the region has industrial biotechnology as a smart specialisation.

Private partners finance intermediate projects (between 15% and 50% of total costs). Public funding can come from individual states or Europe under Horizon 2020.

Source: Courtesy of Michael Manach, TWB.

References

BBI (n.d.), "Bio-based Industries: Public-private Partnership", webpage, https://bbi-europe.eu (accessed 7 December 2017).

Carrez, D. (2016), "The current status of the BBI Joint Undertaking", presentation at the OECD workshop *Innovation for a Sustainable Bioeconomy*, Paris, 25 May.

Carrez, D. (2014), "The bio-based industries initiative PPP and a focus on higher added-value", presentation at the "Present and future policy for bio-based production" workshop, Turin, 9 October.

Coetzee, W. (2016), "Sweet dreams", *The Chemical Engineer*, Vol. 899, Institution of Chemical Engineers, Rugby, United Kingdom, pp. 30-33.

EC (2016), "Road map: Energy Union Integrated Strategy on Research, Innovation and Competitiveness", http://ec.europa.eu/smart-regulation/roadmaps/docs/2016_rtd_001_energy_union_research_strategy_en.pdf.

EIB (n.d.), "InnovFin – EU Finance for Innovators", webpage, www.eib.org/products/blending/innovfin (accessed 7 December 2017).

Fernández Gutiérrez, M. (2016), "Bio-based industries in the European bioeconomy", presentation at the "Innovation for a sustainable bioeconomy" workshop, Paris, 25 May.

Finland Times (2015), "Spinnova's project wins bio-refinery competition," *Finland Times*, 26 February, www.finlandtimes.fi/business/2015/02/26/14606/Spinnova%E2%80%99s-project-wins-bio-refinery-competition.

House of Commons (2011), "The Green Investment Bank: Second Report of Session 2010-11", *Report* No. HC505, 11 March 2011, The Stationery Office, London.

ICMA (n.d.), "Green Bond Principles", webpage, www.icmagroup.org/Regulatory-Policy-and-Market-Practice/green-social-and-sustainability-bonds/green-bond-principles-gbp (accessed 7 December 2017).

Il Bioeconomista (28 April 2015), "EU Parliament supports shift towards advanced biofuels", Il Bioeconomista blog, https://ilbioeconomista.com/2015/04/28/eu-parliament-supports-shift-towards-advanced-biofuels.

McManis, C.R. and S. Noh (2006), "The Impact of the Bayh-Dole Act on genetic research and development: Evaluating the arguments and empirical evidence to date", 13 August, www.law.berkeley.edu/files/mcmanis(1).doc.

Milken Institute (2013), *Unleashing the Power of the Bio-economy*, Milken Institute, Washington, DC.

OECD (2016a), "3rd OECD Green Investment Financing Forum", webpage, www.oecd.org/environment/cc/2016-green-investment-financing-forum.htm (accessed 7 December 2017).

OECD (2016b), *Green Investment Banks: Scaling up Private Investment in Low-carbon, Climate-resilient Infrastructure*, Green Finance and Investment Series, OECD Publishing, Paris, http://dx.doi.org/10.1787/9789264245129-en.

OECD (2015), "Genomics for sustainable development in emerging economies: Food, environment, and industry", Report of the joint HUGO-OECD workshop, 14 March, Kuala Lumpur, OECD, Paris.

OECD (2003), *Entrepreneurship and Local Economic Development Programme and Policy Recommendations*, OECD Publishing, Paris, http://dx.doi.org/10.1787/9789264199798-en.

Panoutsou, C. (2015), "Integrated biorefineries and innovations in the optimal use of biomass", presentation at the workshop on financing possibilities for key enabling technologies and digital transformation projects under the investment plan, Brussels, 10 December.

Peplow, M. (11 March 2014), "Cellulosic ethanol fights for life: Pioneering biofuel producers hope that US government largesse will ease their way into a tough market", Nature News and Comment, Vol. 507/7491, pp. 152.

Scarlat, N. et al. (2015), "The role of biomass and bioenergy in a future bioeconomy: Policies and facts", *Environmental Development*, Vol. 15, Elsevier, Amsterdam, pp. 3-34.

Suurs, R. and E. Roelofs (2014), "Biobased investment climate in the Netherlands and Europe. Summary results quick scan", www.tno.nl/media/3387/quickscan_biobased_investment_climate_executive_summary.pdf.

VTT Ventures Ltd. (2015), "Spinnova developing environmentally friendly yarn thread technology based on spruce and pine fibres", VTT Ventures Ltd., 31 March, www.vttresearch.com/media/news/spinnova-developing-environmentally-friendly-yarn-thread-technology-based-on-spruce-and-pine-fibres.

Chapter 7.

Biowaste biorefining

Vast tonnages of organic waste materials are available worldwide, which seems to circumvent concerns about using food crops as feedstocks for biorefining. The idea of using organic waste is consistent with other major policy goals, especially a circular economy, which minimises waste generation and promotes a greater level of recycling in society. Biorefining of such "biowastes" goes further: it takes materials that are effectively worthless and turns them into value-added products. But are these materials really waste? What of municipal waste as a feedstock? Is the completely rural setting the optimum location, or does a coastal-rural location make more sense when agriculture is out-of-season? This chapter explores such questions, as well as the potential for public policy clashes.

Introduction

The term "waste" (Box 7.1) as related to use as feedstock in biorefineries refers to a wide range of materials. They include: agricultural residues, such as straw and animal manure and sludges; by-products of animal rendering, especially animal fat; forestry residues; waste industrial gases, especially carbon monoxide (CO) and carbon dioxide (CO_2); and the organic fraction of municipal solid waste (MSW), such as food wastes and plastic waste if not sorted for recycling. Nevertheless, waste biorefining will need, on a case-by-case basis, to be investigated regarding its true sustainability. For example, the collection of waste materials and their delivery to a biorefinery site has both economic and environmental costs. These involve the use of fossil fuels and concomitant greenhouse gas (GHG) emissions for their transportation. Careful supply chain design and security will be essential.

It is important to distinguish between different levels of waste when designing supply chains for biorefineries. Materials like straw, for example, may not be waste materials at all. They could have other uses such as wheat and barley straw for animal bedding. Indeed, calculating the volumes of such materials could be part of a biorefinery roadmap (national or regional). Ideally, since agricultural wastes are seasonal, a waste biorefinery should be able to process multiple waste streams; forestry residues may not be readily available in winter months, and municipal waste should be available year-round.

Box 7.1. Waste or resource?

It is fashionable to use the word "resource" to describe waste since, in theory, all waste should be a resource to achieve the circular economy. "Resource" might be used in the context of a feedstock such as sugar, or sugar cane. On the other hand, bagasse is a fibrous "waste" material of sugar cane processing that can also be used in biorefining; it too is arguably a resource. Further, materials that end up in landfill sites, or are burned or similarly discarded, will be termed "waste". Wood chips are manufactured products used for bioenergy purposes. However, forestry residues, for example, are "waste" materials of forestry that can eventually become a resource. Wastes could alternatively be considered "renewable resources" that can be used and reused to generate valuable and marketable products (Velis, 2015). A description that would avoid conflict would be "secondary raw material feedstock".

The EU Waste Framework Directive defines waste as any substance or object that the holder discards or intends to discard or is required to discard.[1] It also sets out the requirement to manage waste in accordance with a "waste hierarchy". The hierarchy affords top priority to waste prevention, followed by preparing for reuse, then recycling, other types of recovery (including energy recovery) and last of all disposal (e.g. landfill). This definition of waste can lead to problems in using such biowastes as feedstocks for biorefining.

1. www.gov.uk/waste-legislation-and-regulations#eu-waste-framework-directive.

The earliest biorefineries in the modern era of industrial biotechnology date effectively from the beginning of the 21st century. They were often ethanol biorefineries, already common in Brazil, that used food crops as the source of biomass to produce fermentable sugars. For the vast majority of countries, the luxury of home-grown, highly efficient, highly sustainable sugar cane as the source of carbon is not possible. The 21st century boom arrived with corn starch biorefining to ethanol for two purposes: as a replacement for methyl tertiary butyl ether (MTBE) as a fuel oxygenate; and as a gasoline supplement (typically a 10% blend of ethanol with 90% gasoline), with a view to further high percentage ethanol fuels (typically E85, with 85% ethanol).

It was not long, however, until controversy arose over use of a food crop for energy purposes. From the early years of this century, many have seen food crops as a biomass source for liquid biofuels production. The bioethanol industry based on corn (maize) as a feedstock (first-generation biofuels) expanded rapidly. This stoked concern over the role of biofuels in food price increases around 2008, the so-called food vs. fuel debate (e.g. Mueller et al., 2011). Evidence links first-generation biofuels to the price spike, some of it showing a marginal effect among a host of factors. However, the actual extent of the linkage will probably never be known. Many studies (e.g. Abbott et al., 2008: Timmer, 2008, IFPRI, 2010; De Gorter et al., 2013) have identified a complex interaction of causes, of which biofuels were only a part. However, the quest was already underway to use organic waste sources as carbon sources in future biorefineries.

Using waste materials in biorefining has several advantages. It relieves pressure on land, thereby enhancing sustainability. It avoids issues both around indirect land-use change (ILUC) (Van Stappen et al., 2011) and the food vs. fuel debate. Through these three actions, it improves public opinion. Further, in the case of waste industrial gases, especially CO and CO_2, it also uses GHGs that would otherwise become emissions. In other words, it contributes to science and policy goals around reducing emissions in climate policy. In the case of MSW, all of the above apply (as MSW is converted to methane in landfill sites, and methane is a much more potent GHG than CO_2). MSW also addresses an additional policy challenge – the diminishing supply of suitable sites for new landfills, a problem for many countries.

Flexible waste management regulation

Overly stringent waste management regulations can disable the exchange of waste materials in industrial symbiosis. For example, some countries would not have approved the piping of flue gas from Statoil to Gyproc at Kalundborg and the sale of liquid sulphur by Statoil to Kemira because both substances would be classified as hazardous waste. Waste regulation has become increasingly stringent in most OECD countries. The Danish waste regulation system, however, is quite flexible; the Danish Ministry of the Environment also encourages industry to find uses for all waste streams on a case-by-case basis. This allows companies to focus on finding creative ways to become more environmentally benign instead of "fighting the regulator" (Desrochers, 2002). In Europe, the legal qualification of some residues or co-products as waste hinders a broad range of potential biorefinery initiatives. Furthermore, local environmental and spatial permits for managing biowastes are limiting possibilities (Fava et al., 2015).

In this context, policy that encourages an institutional framework that forces companies to internalise their externalities should be given high priority. Such a policy should leave companies the necessary freedom to develop new and profitable uses for by-products.

Geography and its importance for public policy

In recent years, much has been said of rural biorefining, an approach that has pros and cons. One policy goal of a bioeconomy, for example, is rural regeneration. This is needed in many OECD countries as agriculture has become more efficient, drastically reducing the proportion of people working in the sector. As the landfill dilemma is principally an issue of large conurbations, however, the rural model for MSW biorefining is less likely to be attractive: there is often public resistance to building landfills in rural locations to take urban waste. It is equally likely this will apply to rural MSW biorefining unless there are significant incentives, such as local jobs.

The landfill dilemma and lessons for waste biorefining

It is becoming more difficult to find suitable sites for properly engineered landfilling in most countries. Even in Australia, with its large land mass and low population, the available supply of landfill is arguably a scarce resource to be used conservatively (Pickin, 2009). In Japan, with its limited space and high population density, it is becoming increasingly difficult to obtain public acceptance for waste disposal facilities, such as landfill sites; there is rising pressure on land use and growing public concern over environmental and health protection (Ishizaka and Tanaka, 2003). Some regions of the United Kingdom are facing the prospect of no easily accessible landfill sites within the next five years (CIWM Journal, 2017).

Since the 1980s, more than three-quarters of all landfills in the United States have closed (Biomass Magazine, 2011), while waste quantities have ballooned. Across the country, waste output has gone up about 65%, with over half still being landfilled (US EPA, 2014). The waste output of Chicago, Illinois, is now more than 300% what it was in the early 1980s, with remaining landfills getting farther from the city. Figures for 2013 show an Illinois-wide landfill life expectancy of 21 years (Illinois Environmental Protection Agency, 2014). For Chicago itself, landfills could last less than ten years. Since 1997, four New York City boroughs have sent MSW by road or rail to landfills as far away as Ohio, Pennsylvania, South Carolina and Virginia. Meanwhile, New York State has imported MSW from New England and Canada to its up-state landfill sites.

In the European Union, the waste management and recycling sector has a high growth rate. In addition, it is labour-intensive, providing between 1.2 million and 1.5 million jobs (Fava et al., 2015). Waste volumes, however, continue to grow. Variation is maximal: some countries landfill 100%, others nil (OECD, 2017a). On the whole, European data show that preferences for treating waste have shifted in the past decade. More waste is being pushed up the waste hierarchy to be recovered for energy or recycled.

Meanwhile, new landfills might be the single least-popular kind of construction for a municipality, with an array of complex regulatory issues. These include siting restrictions in floodplains, wetlands and faults, as well as the need to protect endangered species, surface water and groundwater. Other considerations include disease and vector (rodents, birds, insects) control; open burning prohibitions; explosive methane gas control; fire prevention through use of cover materials; prevention of bird hazards to aircraft; and closure and post-closure requirements. Thus, from several directions, there is continuous pressure to reduce the amount of material being landfilled. Some MSW, if it can be sorted, can be directed towards biorefining.

Furthermore, there are powerful policy motivators against new landfills. For example, in the European Union, the “landfill directive” – Directive 99/31/EC – limits the quantities of biodegradable wastes (kitchen and similar wastes, including paper) that can be landfilled. Sending organic material to landfill can then be discouraged via taxes on landfill tipping (Scharff, 2014). Several US states, including Connecticut, Vermont, California and Massachusetts, are passing legislation to drive organic waste diversion. This policy (slowly) creates regulatory pressure to adopt other conversion technologies. Over the last decade, Japan has shifted from a waste management policy to an integrated waste and material management approach that promotes dematerialisation and resource efficiency. Landfill shortage and dependency on natural resources imports have been key drivers of these changes (OECD, 2010).

Alternative models to consider

Figure 7.1 examines some of the local geographical, infrastructure and social conditions that must be considered to develop alternatives to rural locations for biorefineries.

Figure 7.1. **Alternatives to the entirely rural model for biorefinery locations**

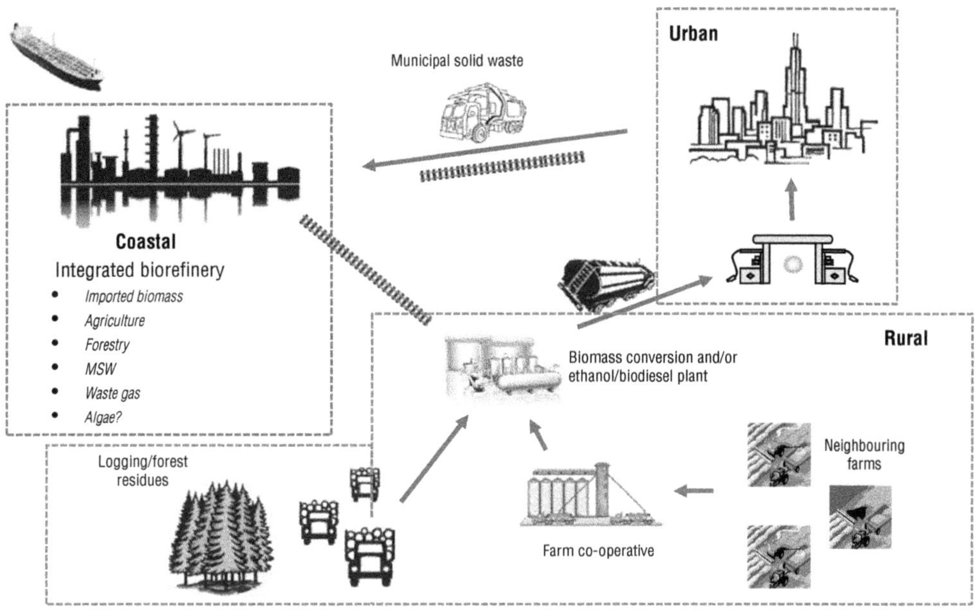

Note: MSW = municipal solid waste.

Source: OECD (2017b), *The Next Production Revolution*, http://dx.doi.org/10.1787/9789264271036-en.

Why the coastal/rural or coastal/suburban biorefinery makes sense

Importing biomass, specifically wood chips, for electricity generation may be necessary or desirable. For this purpose, a coastal location with port facilities makes sense. However, it may not make sense to transport wood chips into the rural setting to generate electricity and then send it back to a city. Many cities struggle to regenerate former industrial sites on coasts such as docklands.

To compensate for the loss of a large biorefinery in the countryside, it may make economic sense to build small industrial facilities in rural locations for several reasons:

- This would bring some jobs to the countryside (rural regeneration).
- Transporting agricultural and forestry residue biomass, low in energy density, does not make economic sense. Converting this biomass into ethanol and/or concentrated sugar solutions or biocoal at rural cellulosic plants may make better sense. (Storing a concentrated sugar solution also provides a biorefinery feedstock outside of the crop growing seasons). Ethanol can then be sent either to the large integrated biorefinery or a petrol blending plant, or both. This creates at least two markets for ethanol – for fuel and for chemicals.
- Many cities struggle to regenerate former coastal industrial sites e.g. docklands.
- Transport distances would be smaller.

- Environmental footprint of the small plant would be less than a full integrated biorefinery, and there would be less conflict with brownfield policies.[1]
- It is still possible in a small facility to generate electricity.
- There could be significant numbers of indirect and induced rural jobs e.g. warehousing, farmers' co-operatives to collect agricultural residues, haulage jobs.
- Small facilities require lower quantities of water – the Crescentino biorefinery, for example, supplies all its water needs from biomass and requires no river water.

It likely takes less time to transport MSW by road, rail or barge over relatively short distances to a coastal location than to a rural facility. Hauling MSW into a rural location could be unpopular with country people (smells, wear-and-tear on roads, safety issues around schools).

Another factor for consideration is the future commercial deployment of marine biorefineries, to date still struggling behind other biorefinery types. Abundant seawater and access to waste CO_2 from, say coastal petro-refineries and petrochemicals plants, may play a major role in determining the location of marine biorefineries. It might be prudent to build integrated biorefineries at coastal locations so that future marine biorefineries could be co-located when ready for deployment.

Waste materials available for bio-based production

Theoretically, a vast treasure trove of solid, liquid and gaseous wastes is available (Figure 7.2), but limited in practice for various reasons. Collecting straw or forestry residues, for example, may not be worthwhile for farmers or forest owners, who thus may need incentives. Municipal solid waste contains a lot of fermentable materials, but they are mixed up with non-fermentable materials. Industrial waste gases exist in profusion and are often in a relatively pure form. However, microbial processes for their fermentation are immature, giving companies little incentive to capture waste gases.

Figure 7.2. **Estimates of lignocellulosic waste materials available globally for bio-production**

Million tonnes

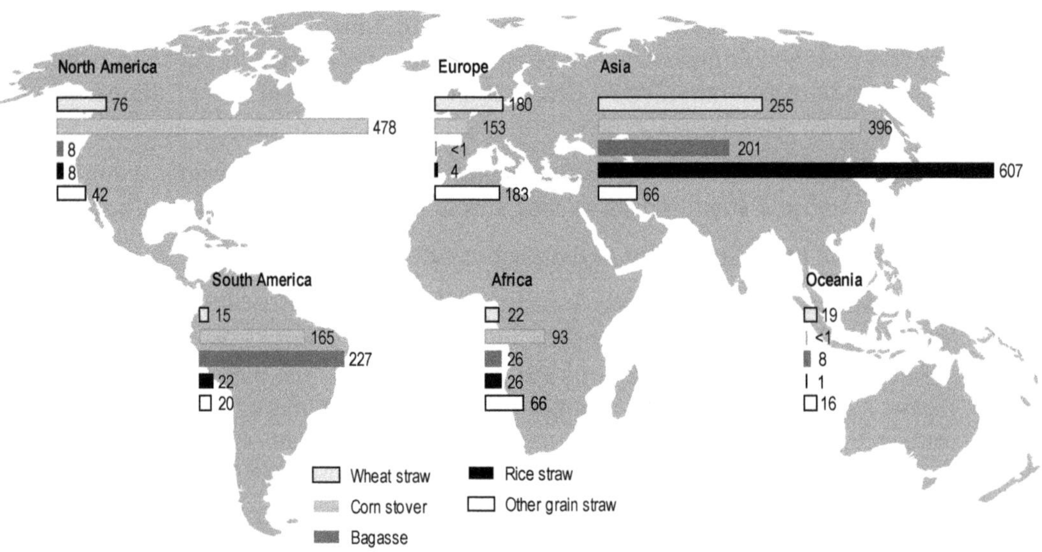

Source: Redrawn from KTN (2016), *From Shale Gas to Biomass: The Future of Chemical Feedstocks*.

A large amount of waste can be used as feedstock, but political will is needed to provide incentives for its collection. In the case of rice straw, for example (OECD, 2015), well over half a billion tonnes is available in Asia, and this material is routinely burned.

Bio-production bottlenecks in the United States have occurred due to multiple factors. These include high costs of both biomass resources, and enzymes or chemicals to break down biomass. Other factors include the recalcitrant nature of lignocellulosic feedstocks and the need for optimised bioprocesses for a wider array of varying feedstocks. The US Department of Agriculture (USDA) has been addressing the need for new feedstocks (Box 7.2), while helping maintain and develop the first-generation ethanol and biodiesel industry.

Box 7.2. The need for new feedstocks in the United States: Initiatives of the USDA

To address bio-production bottleneck factors, the US Department of Agriculture (USDA) introduced five Regional Biomass Research Centers. As one advantage, this programme provided incentives for field researchers (those optimising crops as feedstocks for biofuels) to work closely with researchers developing biorefinery technologies. As the industry evolved, focus has gone from creating corn- and grain-derived ethanol to creating cellulosic ethanol. It is moving towards integrated processes that produce drop-in replacement to petroleum products. Technologies to produce advanced biofuels such as *n*-butanol, pyrolysis bio-oil, hydroxymethylfurfural, liquefied biogas and even (bio)hydrogen have been developed and are arguably commercially viable.

Still, the corn ethanol industry is a multi-million dollar enterprise that merits research towards making it as efficient as possible. One strategy to reach the Renewable Fuels Standard (RFS) targets is to make stepwise improvements in the existing biorefinery concepts. These stepwise improvements must include a regional strategy that builds in enough flexibility to use the "cheapest sources of renewable carbon" within a given region. Such flexibility implies, for example, using grain sorghum, switchgrass or miscanthus in the US Midwest; sweet sorghum or cane sugar in the US South; guayule bagasse in the US Southwest; almond hull sugars in California; and even citrus peel waste in Florida. Another key element is the ability to integrate existing ethanol plants into other operations. Specifically, this enables thermochemical conversion of all biomass sources or integrated digesters to produce biogas and biogas-derived products. Biorefinery strategies are best optimised when field feedstock research on yield, crop quality and biomass cost are co-ordinated with biorefinery strategies (Orts and McMahan, 2016).

Source: Courtesy of Harry Baumes, USDA.

Waste gases

Adani (2015) has attempted to quantify how much waste from different categories is available and to put those numbers into the context of industrial production. Fermentable gases are produced in large quantities from different sectors. However, their collection from some of these sectors is not feasible. Two that are feasible for collection also contribute significantly to emissions: energy supply and industry.

Clearly, in the sectors where collection is feasible, CO_2 is by far the most important gas, although methane (CH_4) is far more potent as a GHG. Four critical figures given by Adani (2015) regarding the potential of gas use in waste biorefining are:

- consumption of renewable raw material for chemical industry and others: 857 million tonnes per year
- total mass used producing chemicals: 271 million tonnes per year

- total mass from CO_2 industry and energy production: 7 596 million tonnes per year
- total mass from biowaste and food loss: ~ 354 million tonnes per year.

The figures suggest, at least on a superficial level, that the amount of CO_2 available far exceeds requirements. Totals, however, can mask many feasibility issues. These include the efficiency of gas use in biorefinery operations, as well as other technical aspects relating to purity of gases, ease and cost of collection. Some preliminary estimates from LanzaTech, a leading company in gas fermentations, suggest that more than 30 billion gallons per year of high-value products can be produced from steel mill waste gases alone; this is a considerable contribution to the worldwide energy and chemical pool (AIChE, 2011).

Residual biomass

Bentsen et al. (2014) suggested more than 3.5 billion tonnes of residual biomass are generated every year in the world, representing about 66% of world energy consumption in transport. In Europe, another study identified 900 million tonnes per year of waste and residues (IEEP et al., 2014). Considering existing competing use and soil quality conservation, 223-225 million tonnes per year of residual biomass are available for advanced biofuel production. This is equivalent to 12% of current road fuel consumption or 16% of projected consumption in 2030.

The UK Department for Environment, Food and Rural Affairs (Defra) estimates that 100 million tonnes of biowaste are available for biogas production in the United Kingdom. This includes agricultural residues, food and drink waste and sewage sludge (House of Lords, 2014). The serious caveat is about purity. Every stage in a bio-based process that requires purification of material represents an additional cost.

The problem of terminology and definitions, and how these influence potential estimates

Figure 7.3 shows several estimates of the quantities of waste materials generated annually in the European Union. There is a problem of definition, which leads to huge variation in figures across different sources.

The figures in (a) and (b), for example, are quite different, which may relate to the difference between "agricultural residues" and "agricultural waste". Comparing (c) with (b), the numbers for "sludge" are also very different. The use of the term "biowaste" in (c) could incorporate all of the categories in (b). The numbers in (d) refer to "waste biomass" in the European Union, 2012.

Therefore, the mixture of terms and a lack of standardised definitions make it difficult to truly assess the volumes of different (waste) materials that can be used in biorefining. Conversely, volumes from crop feedstocks (e.g. sugar cane or sugar beet) are collected internationally and readily comparable. Therefore, an important message for both the public and private sectors is the need for standard terms and definitions. For the public sector, standards are important when attempting to make strategic documents like biorefinery roadmaps. For example, how would it be possible to create a timeline for a national or regional biorefining industry in the absence of certainty around feedstock volumes? For the private sector, building a biorefinery to a certain tonnage capacity also needs certainty on available feedstocks.

Figure 7.3. **Data from different sources highlight the discrepancies in waste potential**

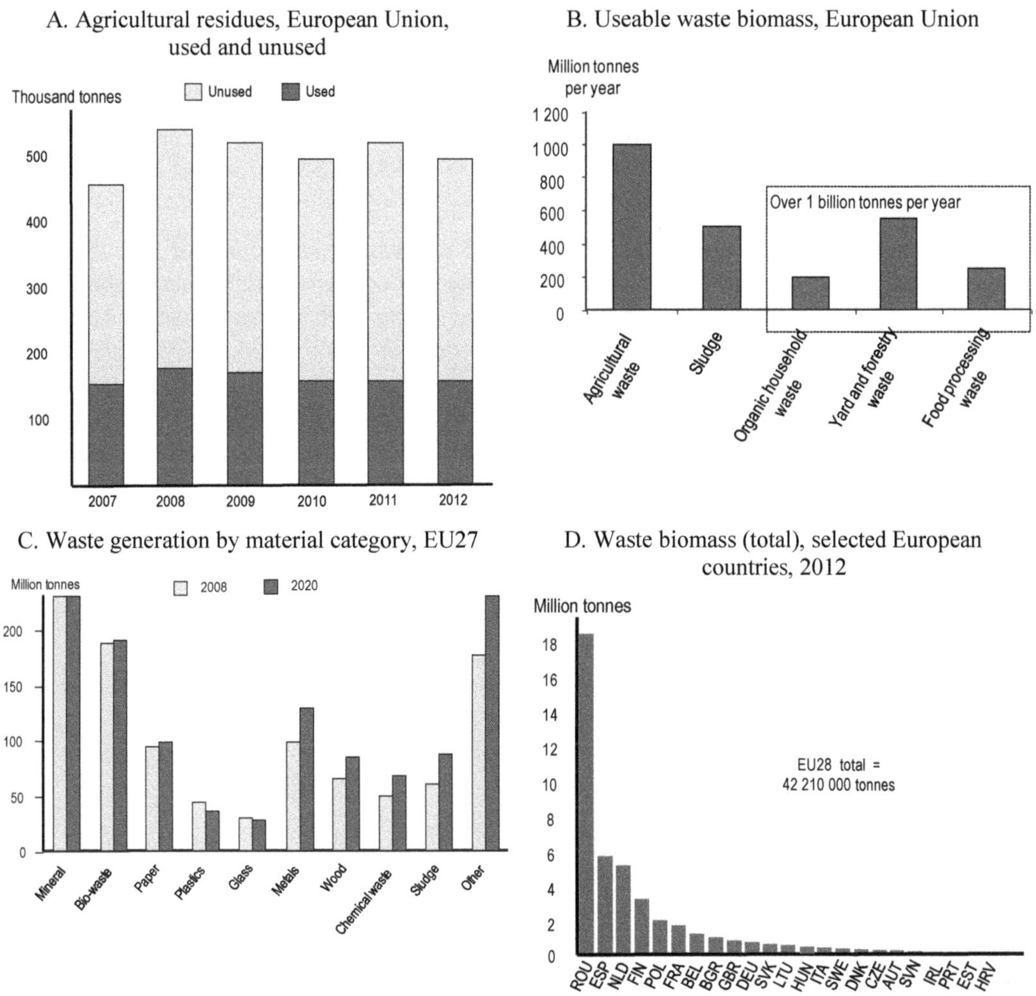

Sources: (a) https://biobs.jrc.ec.europa.eu/market/agricultural-biomass (accessed in 2016); (b) Fava et al. (2015), "Biowaste biorefinery in Europe: Opportunities and research & development needs"; (c) OECD (2014), "Present and future policy for bio-based production"; (d) https://biobs.jrc.ec.europa.eu/market/waste-biomass-total (accessed in 2016).

The development of common definitions will enable better data collection by both private and public entities. This would help resolve the issue of comparison between different data sources mentioned above.

- "Bioeconomy": lack of an agreed definition is a hindrance (denies the science input, no international databases, possible trade barriers).
- "Biowaste": most statistics do not distinguish between wet and dry weight, so no comparisons can be performed. It is extremely important to clarify the definition of biowaste. According to the European Commission:

> Biowaste is defined as biodegradable garden and park waste, food and kitchen waste from households, restaurants, caterers and retail premises, and comparable waste from food processing plants. It does not include forestry or agricultural residues, manure, sewage sludge, or other biodegradable waste such as natural textiles, paper or processed wood. It also excludes those by-products of food production that never become waste. (EC, 2018).

By leaving out forestry and agricultural residues, the tonnages generated will be very different.

- The definition of "waste disposal" could be changed to allow collection, transportation and sorting in view of its conversion in biorefineries. If a material is to be converted in a biorefinery then it should effectively no longer be regarded as a waste, but as a resource. If this is done officially, it will nullify many problems around collection and transport.
- A definition of "bio-based product" and a harmonised framework for bio-based products are needed as a standard for public procurement and business development. The European Committee for Standardisation (CEN) has made progress in development of such a framework, but there is still a need to spread use of the developed standards to capitalise on their market pull potential. This international co-operation can be done by, for example, exchange of best practices and experiences to reach a more coherent approach to bio-based products globally. Without it, trade barriers are certain to develop.
- "Competitive potential", which generally requires an economic model of competing technologies, needs to be assessed. For example, the future of zero-carbon transportation depends on whether cellulosic ethanol becomes economical at large scale and can compete with electric vehicles.

Ultimately, integration of actors across sectors and hence the creation of new value chains is limited by disparity, as well as lack of both control of terminology and standards. In short, a commonly agreed vocabulary throughout value chains is needed – from feedstock suppliers to biorefining to downstream actors in the application sectors.

Municipal solid waste volumes

> CEO [of Enerkem] Vincent Chornet looked at the big picture of potential, and it is big. Although there are 1.3 billion metric tonnes of MSW, about 420 million of them are suitable for Enerkem. That's as much as 160 billion liters (42 billion gallons) of renewable fuels (or chemicals) from one sector alone – more than doubling the addressable market for biofuels with just the one feedstock – and vastly outstripping the current [dollars] being brought in via waste to energy (incineration) technologies, which is around $7.6B, or a fraction of the $70B+ market available with the new technology. (Lane, 2015b.)

The figures for tonnages of MSW (Box 7.3) mentioned above are global tonnages. The figures merit further investigation from the public policy perspective. Although this appears to be an unprecedented opportunity to really make a difference to the landfill dilemma, the potential interaction between the private sector and public policy must be examined. For example, would this activity interfere with other markets, especially recycling, energy recovery and electricity generation, and industrial composting?

Addressing the latter part of the quotation, combusting mixed waste also comes with issues. These include cost, sorting, scrubbing the gas stream to remove toxins, GHG emissions, and, in some locations, negative public reaction. Moreover, as the quotation hints, the product – electric power – is low value and effectively zero value added.

Different figures give a perspective on what MSW tonnages translate to in bio-based production (Table 7.1).

Table 7.1. **Conversion of tonnages of MSW into crude oil and bio-based equivalents**

Quantity of MSW = 260 million tonnes/year

Biomass feedstock (10% water)	140 400 000 tonnes per year
Crude oil equivalent	322 436 000 barrels per year
Diesel fuel equivalent	14 490 billion gallons per year
Ethanol equivalent	24 500 billion gallons per year
Electricity equivalent	164 300 000 megawatts per year

Source: Hennessey (2011), "Biomass feedstock from MSW: Backbone for the biorefining industry".

Box 7.3. **What is municipal solid waste?**

Generally, in European countries and OECD countries, municipal solid waste (MSW) covers waste from households (82% of total MSW), including bulky waste. The remainder of MSW comes from commerce and trade, office buildings, institutions and small businesses, yard and garden waste, street sweepings, the contents of litter containers and market cleansing waste (Eurostat, 2003). The definition of MSW excludes waste from municipal sewage networks and treatment, as well as municipal construction and demolition waste. However, national definitions of MSW may differ (OECD, 2007). In a developing economy, MSW is generally defined as the waste produced in a municipality. Most MSW generated in developing countries is non-segregated and, therefore, either hazardous or non-hazardous (Karak et al., 2012). Many countries likely contain a significant amount of food waste, which is extremely useful for gasification or fermentation.

About 65% of municipal waste is biodegradable. The EU Directive on the landfill of waste aims to reduce environmental pressures from landfill, particularly methane emissions and leachates (Official Journal of the European Communities, 1999). It requires member states to reduce landfill of biodegradable municipal waste to 75% of the amounts generated in 1995 by 2006, to 50% by 2009 and to 35% by 2016.

In the United States, the number of landfill sites has dropped by 75% in the past 25 years. However, this number is deceptive. Much of the decrease is due to consolidation of multiple landfills into a single, more efficient facility. Also, technology has allowed for each acre of landfill to take 30% more waste. So, during this time, the available landfill per person has actually increased by almost 30%. As of 2010, total US MSW generation was 250 million tonnes. Paper and paperboard account for 29%, and yard trimmings and food scraps account for another 27%. The rest breaks down as follows: plastics 12%; metals 9%, rubber, leather and textiles 8%; wood approximately 6.4% and glass 5% (Hennessey, 2011).

The earliest MSW biorefineries are open for business

At least two high-profile biorefineries have been established through public-private partnerships to convert MSW into bioethanol and methanol. The facility in Ineos Vero Beach, Florida, which received a USD 75 million loan in 2011 (USDA, 2014), is relatively small. In 2013, it began producing 8 million gallons of cellulosic ethanol per year from vegetative and yard waste, as well as MSW. The other is the Enerkem plant in Edmonton, Canada. Both are gasification and fermentation plants i.e. gasification is needed to get MSW ready for use as a feedstock.

Is MSW biorefining a truly sustainable and economic business model?

In the face of growing waste management and disposal costs, the demand for petro-based products – fuel, plastics or chemicals – also continues to rise. Although governments have been notoriously slow to adopt sustainability policies, sustainability goals and mission statements are increasingly common among many large corporations. Indeed, in the absence of public policy, industry may go it alone. However, this may not result in the most sustainable solutions or the most desired public policy goals.

The policy pros and cons

This section is largely a summary and extrapolation of some considerations in RWI (2014).

There are two potential revenue streams for a biorefinery facility, which are both uncertain: the gate or tipping fees[2] from taking the waste; and revenues from selling biofuels. Gate fees vary enormously by country and region, and landfill tax tends to make gate fees higher. Where gate fees are low, the production of biofuels from waste is not cost-competitive with landfill. Therefore, public stimulus is needed for countries, regions or cities to break out of the landfill dilemma.

For waste treatment facilities such as incinerators or composting plants, the fee offsets the operation, maintenance, labour costs and capital costs of the facility along with any profits and final disposal costs of any unusable residues.

For some years, many have argued for a policy shift to offer more support for bio-based chemicals. In this particular case, chemicals usually have higher margins than liquid fuels, have more value added and create more jobs than biofuels. Therefore, diversifying MSW biorefineries so they can also make bio-based chemicals would seem to improve the economics irrespective of gate fees.

This is a competitive market. Anaerobic digestion (AD) is a tried-and-tested technology that has been brought up-to-date in the last decade; it now involves the anaerobic fermentation of waste to biogas, which is over 50% methane. AD facilities are generally cheaper to design and build than waste-to-biofuels biorefineries, plus they are significantly better proven. The flexibility of AD as a process allows for biogas to be used to generate electricity. It can be piped as gas and create fertiliser, and be adapted to provide combined heat and power.

Incineration is also both proven and effective at disposal and energy generation. Early incinerators had a bad reputation, but the challenges have been overcome. In Japan, incineration with energy capture has been increasingly popular as it can be used to tackle the vast waste plastics problem (Yamashita and Matsumoto, 2014).[3] Burning the other organic fraction of MSW with plastics reduces the sorting difficulties. In effect, MSW biorefineries are in competition with other buyers such as incineration utilities (Knight et al., 2015).

There are counter-arguments that favour waste-to-biofuels (and/or chemicals). First, the technology creates fuel from non-recyclable and non-compostable MSW i.e. it can work in partnership with other sustainable waste technologies, not against them. Second, more experience is being gained with gasification technology, which will help with the economics and the confidence in using a process such as Enerkem. There is also an embryonic technology to turn waste gases (and natural gas) into animal feed and value-added chemicals through fermentation. Calysta of Norway uses natural gas-fed fermentation to produce feed-quality protein with high nutritional value for use in aquaculture (Calysta, n.d.).

Eventually, the diversity of chemicals that can be produced after gasification will be higher. Environmental regulations are constantly becoming more stringent. Therefore, any technology that can improve both economic and environmental outcomes while creating jobs must be taken seriously, even if alternatives such as landfill are more competitive. Landfill is no solution for the 21st century.

Scale-up is now the critical issue

MSW biorefineries are thus far unproven at commercial scale. Second-generation biofuels are too recent for a long-term success story that could provide evidence of a scalable, repeatable business model. The successes of first-generation ethanol in Brazil are not transferrable to other countries. Thus, there is even less experience with waste-to-biofuels projects and facilities. Without high quality, robust data from functioning operations, the justification for large capital injections will remain a barrier. However, the number of such projects is gradually growing. They can be regarded as flagship projects; if successful, they should help de-risk future projects. Nevertheless, policy makers will be obliged to study the business case carefully on an individual basis. This will require close communication between municipalities and their waste management operators, the private sector and the potential investors along with public agencies offering investment.

Notes

1. In town planning, brownfield land is an area of land previously used or built upon, as opposed to greenfield land, which has never been built upon. Brownfield status is a legal designation that places restrictions, conditions or incentives on redevelopment.
2. A gate fee and tipping fee mean the same thing. It is the charge levied upon a given quantity of waste received at a waste processing facility. In the case of a landfill, the fee is generally levied to offset the cost of opening, maintaining and eventually closing the site. It may also include any landfill tax that is applicable locally. See http://en.wikipedia.org/wiki/Gate_fee.
3. The ultimate destination for about 3% of plastic waste is the oceans. It has been estimated that the plastic waste entering the world's oceans could double in the next ten years (Jambeck et al., 2015).

References

Abbott, P. et al. (2008), "What's driving food prices?", *Farm Foundation Issue Report*, Farm Foundation, Oak Brook, United States.

Adani, F. (2015), "Availability of bio-waste types (including gases)", presentation at Biorefining: bio-waste and organic by-products exploitation in multi-purpose biorefinery schemes workshop, Economdo 2015, 4 November, Rimini Fiera, Italy.

AIChE (2011), "Gas Fermentation of Industrial Waste Gases to Fuels and Chemicals", webpage, https://www.aiche.org/academy/videos/conference-presentations/gas-fermentation-industrial-waste-gases-fuel-and-chemicals (accessed 7 December 2017).

Bentsen, N.S. et al. (2014), "Agricultural residue production and potentials for energy and materials services", *Progress in Energy and Combustion Science*, Vol. 40, Elsevier, Amsterdam, pp. 59-73.

Biomass Magazine (2011), "The role of MSW for future biorefineries", *Biomass Magazine*, 18 January, http://biomassmagazine.com/articles/6746/the-role-of-msw-for-future-biorefineries/?ref=brm.

Calysta (n.d.), "Nutrition", webpage, http://calysta.com/nutrition (accessed 7 December 2017).

CIWM Journal (2017), "UK Faces "Serious" Waste Infrastructure Capacity Gap – SUEZ", *CIWM Journal*, 16 August, https://ciwm-journal.co.uk/uk-faces-serious-waste-infrastructure-capacity-gap-suez.

De Gorter, H. et al. (2013), "How biofuels policies affect the level of grains and oilseed prices: Theory, models and evidence", *Global Food Security*, Vol. 2/2, Elsevier, Amsterdam, pp. 82-88.

Desrochers, P. (2002), "Cities and industrial symbiosis. Some historical perspectives and policy implications", *Journal of Industrial Ecology*, Vol. 4, pp. 29-44.

EC (European Commission) (2018), "Biodegradable waste", web page, http://ec.europa.eu/environment/waste/compost/ (accessed 2 March 2018).

Eurostat, (2003), *Waste Generated and Treated in Europe*, Office for Official Publications of the European Communities, Luxembourg.

Fava, F. et al. (2015), "Biowaste biorefinery in Europe: Opportunities and research & development needs", *New Biotechnology*, Vol. 32/1, Elsevier, Amsterdam, pp. 100-108.

Hennessey, E. (2011), "Biomass feedstock from MSW: Backbone for the biorefining industry", presentation at the municipal solid waste to biofuels 2011 summit, Chicago, 10-11 February.

House of Lords (2014), "Waste or resource? Stimulating a bioeconomy", *Report of the Science and Technology Select Committee*, HL Paper No. 141, The Stationery Office Limited, London.

IEEP et al. (2014), *Wasted: Europe's Untapped Resource: An Assessment of Advanced Biofuels from Wastes & Residues*, Institute for Environmental Policy et al., Brussels, http://europeanclimate.org/wp-content/uploads/2014/02/WASTED-final.pdf.

IFPRI (2010), *Reflections on the Global Food Crisis: How did it Happen? How has it Hurt? And How can We Prevent the Next One?*, International Food Policy Research Institute, Washington, DC.

Illinois Environmental Protection Agency (2014), *Illinois Landfill Projections of Disposal Capacity as of January 1, 2014*, Illinois Environmental Protection Agency, Springfield, United States, www.epa.state.il.us/land/landfill-capacity/2013/landfill-capacity-report-2014.pdf.

Ishizaka, K. and M. Tanaka (2003), "Resolving public conflict in site selection process – a risk communication approach", *Waste Management*, Vol. 23/5, Elsevier, Amsterdam, pp. 385-396.

Jambeck, J.R. et al. (2015), "Plastic waste inputs from land into the ocean", *Science*, Vol. 347/6223, American Association for the Advancement of Science, Washington, DC, pp. 768-771.

Karak, T. et al. (2012), "Municipal solid waste generation, composition, and management: The world scenario", *Critical Reviews in Environmental Science and Technology*, Vol. 42/15, Taylor & France Online, pp. 1509-1630.

Knight, L. et al. (2015), "Supply market uncertainty: Exploring consequences and responses within sustainability transitions", *Journal of Purchasing & Supply Management*, Vol. 21/3, Elsevier, Amsterdam, pp. 167-177.

KTN (2016), *From Shale Gas to Biomass: The Future of Chemical Feedstocks*, Knowledge Transfer Network, Horsham, United Kingdom.

Lane, J. (2015), "Who said what: The hottest slides from ABLC 2015", *Biofuels Digest*, 15 March 2015.

Mueller, S.A. et al. (2011), "Impact of biofuel production and other supply and demand factors on food price increases in 2008", *Biomass and Bioenergy*, Vol. 35/5, Elsevier, Amsterdam, pp. 1623-1632.

OECD (2017a), "Municipal waste generation and treatment", *OECD.Stat*, OECD, Paris, https://stats.oecd.org/Index.aspx?DataSetCode=MUNW (accessed 1 October 2017).

OECD (2017b), *The Next Production Revolution*, OECD Publishing, Paris, http://dx.doi.org/10.1787/9789264271036-en.

OECD (2015), "Genomics for sustainable development in emerging economies: Food, environment and industry", *Report of the Joint HUGO-OECD Workshop*, 14 March, Kuala Lumpur, OECD, Paris.

OECD (2014), "Present and future policy for bio-based production", *Report of an OECD Workshop*, Turin, 9 October 2014, OECD, Paris.

OECD (2010), *Environmental Performance Reviews: Japan 2010*, OECD Publishing, Paris, http://dx.doi.org/10.1787/9789264087873-en.

OECD (2007), *Environmental Data, Compendium 2006-2008*, OECD Publishing, Paris, www.oecd.org/env/indicators-modelling-outlooks/oecdenvironmentaldatacompendium.htm.

Official Journal of the European Communities (1999), "Council Directive 1999/31/EC of 26 April 1999 on the landfill of waste", http://eur-lex.europa.eu/legal-content/EN/TXT/PDF/?uri=CELEX:31999L0031&from=EN (accessed 7 December 2017).

Orts, W.J. and C.M. McMahan (2016), "Biorefinery developments for advanced biofuels from a sustainable array of biomass feedstocks: Survey of recent biomass conversion research from Agricultural Research Service", *BioEnergy Research*, Vol. 9/2, Springer, pp. 430-446.

Pickin, J. (2009), "Australian landfill capacities into the future", report commissioned by the Department of the Environment, Water, Heritage and the Arts, Hyder Consulting Pty Ltd., Melbourne, Australia.

RWI (2014), "Is the conversion of municipal solid waste to biofuels and bio-products a truly sustainable and economic business model in the USA?", *An Analysis of the USA Market and the State of Play for MSW to Biofuels and Bio-products Projects*, Renewable Waste Intelligence, London.

Scharff, H. (2014), "Landfill reduction experience in The Netherlands", *Waste Management*, Vol. 34, Elsevier, Amsterdam, pp. 2218-2224.

Timmer, P.C. (2008), "Causes of High Food Price plus Technical Annexes", *Economics Working Paper Series*, No. 128 (October), Asian Development Bank, Manila.

USDA (2014), "USDA announces loan guarantee to help innovative company turn waste into renewable jet fuel", News Release, USDA Office of Communications Release No. 0195.14, US Department of Agriculture, Washington, DC.

US EPA (2014), "Municipal solid waste generation, recycling, and disposal in the United States: Facts and figures for 2012", EPA-530-F-14-001, February 2014, US Environmental Protection Agency, Washington, DC.

Van Stappen, F. et al. (2011), "Direct and indirect land use changes issues in European sustainability initiatives: State-of-the-art, open issues and future developments", *Biomass and Bioenergy,* Vol. 35, Elsevier, Amsterdam, pp. 4824-4834.

Velis, C.A. (2015), "Circular economy and global secondary material supply chains", *Waste Management Research*, Vol. 33/5, Sage Publications, Thousand Oaks, United States, pp. 389-391.

Yamashita, M. and S.-I. Matsumoto (2014), "Status of recycling plastic bottles in Japan and a comparison of the energy costs of different recycling methods", *International Journal of Environmental Protection and Policy*, Vol. 2/4, Science Publishing Group, New York, pp. 132-137.

Part III.

Towards bio-production of materials: Replacing the oil barrel

Chapter 8.

Developments in bio-based production

Much innovation has been achieved in biorefining in the last few years, often in the absence of significant policy support. This chapter highlights the potential of biorefining to replace fossil-derived manufacturing in terms of materials that can be produced. While the examples demonstrate that a wide variety of materials is already available in the market, the chapter also gives a sense of perspective: the real test for the future of bio-production in manufacturing is its ability to produce all these promising materials at a scale appropriate to society.

Introduction

Since the publication *Future Prospects for Industrial Biotechnology* (OECD, 2011), interesting bio-based materials have proliferated (Table 8.3). Promoting chemistry at a political level poses challenges as chemicals are largely invisible, and yet they play an essential role in virtually all manufactured goods. This is compounded by the challenge of an industry that struggles with a poor public image (Moreau, 2005). This chapter will examine some of these new bio-based materials spanning a range of different product types and bringing a visibility not seen before. It will also illustrate the range of different bio-based chemicals that are close to commercialisation.

What would be involved in replacing the oil barrel?

The day could come when light and medium transport can be electrified (Delucchi et al., 2014), thereby eliminating the need for liquid road transport fuels. For example, Scania of Sweden is introducing a hybrid truck for city use that can be driven electric-only or with renewable fuels (Scania, n.d.). For its part, Tesla unveiled an all-electric truck in late-2017. The Swedish government aims to have a fossil-independent vehicle fleet by the year 2030 (Hellsmark et al., 2016). France and the United Kingdom declared in mid-2017 that they will be rid of new petrol and diesel cars by 2040.

For shipping and aviation, alternatives to liquid fuels are hard to envisage. Aviation is responsible for up to 3% of global human-made CO_2 emissions. Unlike other forms of transportation, aviation has fewer green alternatives to significantly reduce its carbon footprint. To this end, Los Angeles and Oslo were the first airports in the world to incorporate biofuel into the regular refuelling process (Il Bioeconomista, 2016). Several airlines are now purchasing bio-aviation fuel e.g. KLM and United Airlines. In May 2016, Cathay Pacific commenced a two-year programme of flights from Toulouse to Hong Kong, People's Republic of China (hereafter "China") using renewable jet fuel. In September 2016, Gevo announced it had entered into a heads of agreement with Deutsche Lufthansa AG to supply up to 8 million gallons per year of alcohol-to-jet fuel (ATJ). SkyNRG is a market leader for sustainable jet fuel, supplying more than 20 carriers across five continents (SkyNRG, n.d.).

Without fuels production, petrochemicals might be much less profitable. In the current model, petroleum refiners would have great difficulty producing chemicals at low cost if demand for gasoline or diesel fuel were radically reduced. The business model for upstream oil companies would be radically different, especially as new sources of oil become more expensive.

However, the high standard of living attained in OECD countries is not imaginable without the vast plethora of chemicals in everyday use. As a simple illustrative example, there would be no smart phone without chemistry, or any telephone at all. As 96% of all manufactured goods require at least one chemical (Milken Institute, 2013), petrochemicals will be clearly much harder to replace than fossil fuels. If coal, crude oil and natural gas were conserved (by ending the practice of burning them as fuels), a ready feedstock of fossil resources would be available for future generations to make petrochemicals. In the short term, however, it is extremely unlikely that fossil fuels will no longer be burned. Therefore, interim and long-term policy solutions need to be pursued.

The chemicals sector is the largest industrial energy user, accounting for about 10% of global final energy use (Broeren et al., 2014). It is also the third largest industrial source of emissions after the iron and steel, and cement sectors (IEA, 2012). As some countries struggle to meet their emissions reduction obligations, it is puzzling that the chemical sector has been relatively ignored in this respect compared to fuels and electricity (Philp, 2015).

Later this century, increasing demand for chemicals and plastics may cause a competition with fuels for available crude oil. Between 1950 and 2011, plastics consumption rose with a compound annual growth rate (CAGR) of 8.6%, and is now close to 300 million tonnes per annum (Figure 8.1). Future growth in plastics consumption is predicted to be about 4% per annum (ANZ Insights, 2012). Since the mid-1980s, the global chemical industry overall has grown by 7% annually. Asia has driven most of the growth in the past 25 years. If trends continue, global chemical markets could grow on average at 3% per annum in the next 20 years (AT Kearney, 2012).

Figure 8.1. **World plastics consumption, 1950-2011**

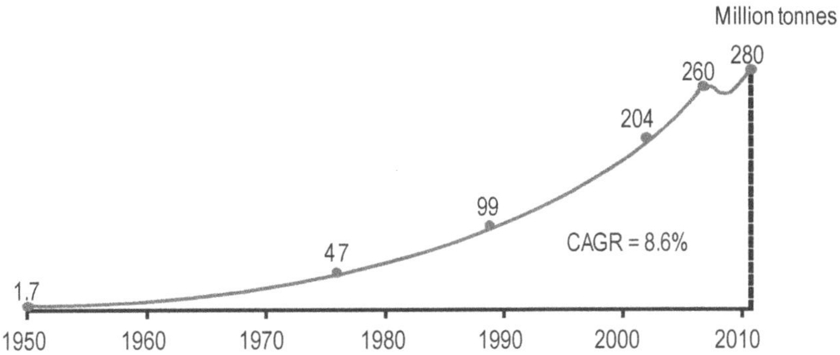

Note: CAGR = compound annual growth rate.

Source: Redrawn from ANZ Insights (2012), "Global plastics industry: Market update".

On that basis, plastics consumption could increase about four-fold by 2050. Approximately 8% of world oil production is used in plastics manufacture: 4% as raw material for plastics and 3-4% as energy for manufacture (Hopewell et al., 2009). Therefore, by mid-century, consumption of crude oil to make plastics could increase to 28-32% of current levels of production. This, in turn, would put plastics in competition with fuels for crude oil. Such growth is completely out of step with new oil discoveries, which are at their lowest in 60 years.

Renewable feedstocks offer the most compelling route to drop-in (exact equivalent) or same-function (different molecule that has the same function) sustainable chemicals. This would previously have been almost entirely the province of chemistry. For example, the whole history of wood chemistry has been largely forgotten since the petrochemicals era (e.g. USDA, 1956), and a lot more can now be done since this early report. More recently, there has been a drive towards "eco-friendly" chemicals, such as the Ecover brand of washing-up liquids. Biotechnology is a relative newcomer as a route to commodity chemicals; it was less than three decades ago that Frost and Lievense (1994) discussed biotechnological routes to aromatics in reference to "environmental considerations and the scarcity of petroleum".

The idea of biotechnological routes to entirely unnatural chemicals only took hold with the emergence of metabolic engineering in the 1990s (Wong, 2016). Many petrochemicals in everyday use have no natural equivalent. They are highly reduced in nature compared to carbohydrates, and often toxic to a microbial catalyst (Yim et al., 2011). This means a daunting task for creating biochemical pathways to a molecule never seen in nature, thus requiring truly synthetic steps. It also requires building other features into a microbial catalyst, such as solvent tolerance. This would create a "robustness" in the microbe, allowing it to survive the conditions of the bioprocess and the toxicity of the desired product.

Despite the challenges, a biotechnological route offers several advantages over a strictly chemical route. Microbial metabolism is extremely diverse, and therefore provides a choice of large numbers of biochemical reactions (one database contains 130 000 hypothetical enzymatic reactions). Biology is often specific and selective, implying that side effects that limit productivity could be minimal or minimised. Microbial processes occur at low temperatures and mostly at ambient pressures, therefore making the biotechnological route attractive in environmental and economic terms.

To date, green, renewable chemistry remains far ahead compared to "renewable biotechnology" in the production of commodity chemicals. Figure 8.2 sums up the challenge.

Figure 8.2. **Chemicals made through metabolic engineering of microorganisms**

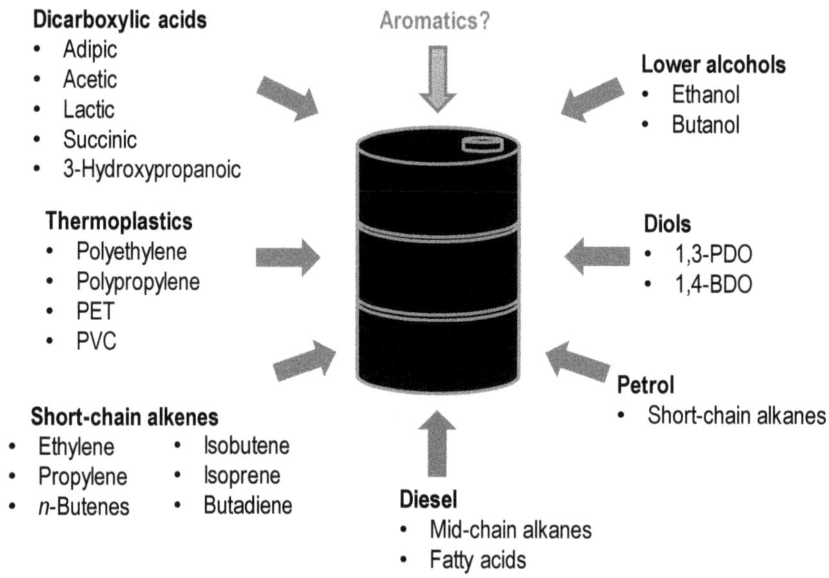

Note: PET = polyethylene terephthalate; PVC = polyvinyl chloride; PDO = propanediol; BDO = butanediol.

Source: Adapted from Jiménez-Sánchez and Philp (2015), "Omics and the bioeconomy: Applications of genomics hold great potential for a future bio-based economy and sustainable development".

Most of the chemicals in Figure 8.2 remain as research successes; many may never reach commercialisation. The technical and financial reasons for this limitation are interlinked. In terms of cost, more efficient biotechnologies would bring down production price and make bio-based (either drop-in or equivalent function) more cost-competitive with petrochemistry. In terms of technical issues, scale-up often significantly reduces performance of engineered strains (e.g. Takors, 2012). Fundamentally, bio-based production without public policy support faces a mountainous challenge given the economies of scale possible in petrochemistry. For example, IRENA/ETSAP (2013) estimated the worldwide production costs of bio-based ethylene to be on average 50% higher compared to the production of ethylene in the steam cracking process.

However, a relatively small number of chemicals represent a large proportion of total organic chemicals production. US DOE (2004) identified 12 building block chemicals that can be produced from sugars via biological or chemical conversions (Table 8.1). Building block chemicals are considered to be molecules with multiple functional groups that can be transformed into new families of useful molecules. They can therefore otherwise be termed "platform chemicals".

Table 8.1. **The US DOE top value-added chemicals from biomass feedstocks**

Chemicals
1,4 diacids (especially succinic, fumaric, malic)
3-hydroxypropionic acid
Levulinic acid
Glutamic acid/MSG
Sorbitol
Xylitol/arabinitol
2,5 furan dicarboxylic acid
Aspartic acid
Glucaric acid
Itaconic acid
3-hydroxybutyrolactone
Glycerol

Note: MSG = monosodium glutamate.

Source: Adapted from US DOE (2004), "Top value-added chemicals from biomass (results of screening for potential candidates from sugars and synthesis gas, Vol. 1)".

Saygin et al. (2014) estimated that seven polymers could *technically* replace half of the total common plastics in use in 2007 (Table 8.2). These polymers were bio-PE, bio-PET, PHA, PTT, PLA, starch polymers and cellulosic films.

Table 8.2. **Top seven polymers (and ethylene) that could technically replace half of total polymers production in 2007**

Material	CO_2 emissions savings (tonnes CO_2 per tonne)
Bio-ethylene	1.9-5.3
Bio-polyethylene (PE)	2.4-4.2
Bio-polyethylene terephthalate (PET)	1.9-2.5
Polyhydroxyalkanoates (PHA)	1.4-4.0
Polytrimethylene terephthalate (PTT)	1.1-1.9
Polylactic acid (PLA)	1.2-2.1
Starch polymers	1.7-3.6
Cellulosic films	0-1.9

Source: Adapted from Saygin et al. (2014), "Assessment of the technical and economic potentials of biomass use for the production of steam, chemicals and polymers".

One significant development has been the arrival of the bio-based equivalents of the major thermoplastics that dominate the market – polyethylene (PE), polypropylene (PP) and polyethylene terephthalate (PET). Bio-PE and bio-PP are produced chemically from monomers that are made through fermentation. They have identical performance characteristics to the petro-based equivalents and, importantly, can directly enter existing recycling systems. They can be categorised as bioplastics as their carbon content comes from renewable resources. As a result, they can make a potential contribution to GHG emissions savings. The global trend in bioplastics production will change significantly, becoming dominated by durable bio-based thermoplastics (OECD, 2013) rather than biodegradable plastics. The most dynamic developments are still expected to be in drop-in bio-based polymers (Aeschelmann et al., 2015).

A question mark exists for the aromatics. Biotechnological routes to aromatics are particularly challenging. As high-volume chemicals with a large range of functions, they cannot easily be replaced: for example, benzene has specific uses in its own right, but has important value chains that lead to even more valuable chemicals. However, commodity aromatics are toxic to microbial cells. Indeed, most microbiological studies with aromatics look at their biodegradation as pollutants rather than their synthesis. Several studies have focused on microbial aromatics production from biomass (Kawaguchi et al., 2016), but not aimed at commodity aromatics.

On the other hand, clear environmental drivers for replacing aromatics exist. The BTX compounds (benzene, toluene and xylene) are mainly produced by catalytic reforming. Typically, this uses hydrogen and catalysts under high temperature (500°C) and high pressure (10-50 bar) (Eriksson, 2013). The largest renewable reservoirs of aromatic materials are lignin and hemicellulose. Lignin creates the greatest challenges for renewable sources of aromatics, but should still not be ignored (Figure 8.3). The total lignin availability in the biosphere exceeds 300 billion tonnes and increases annually by around 20 billion tonnes (Smolarski, 2012).

Figure 8.3. **The potential for renewable aromatics production from lignin**

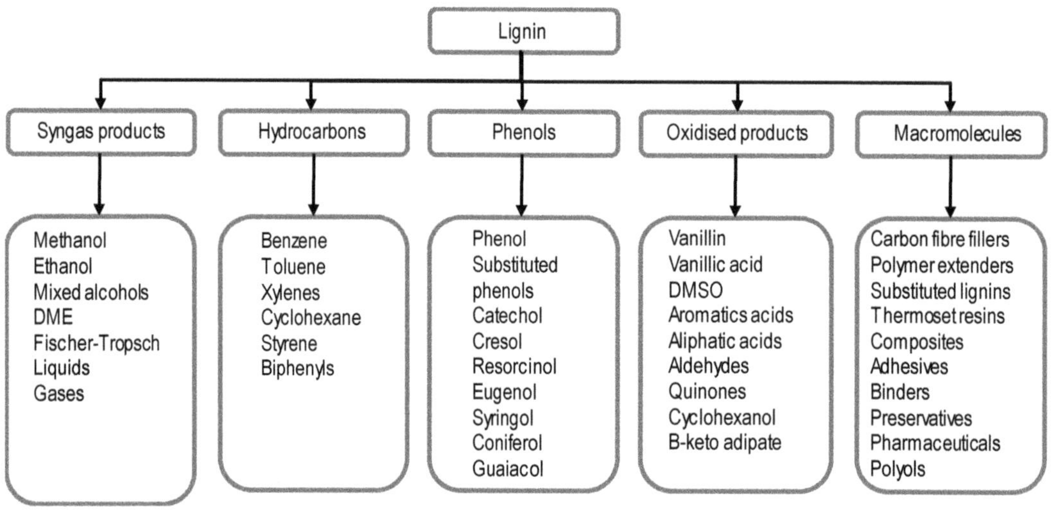

Note: DMSO = dimethyl sulfoxide.

Source: Redrawn from IEA Bioenergy Task 42 Biorefinery (2012), "Bio-based chemicals. Value added products from biorefineries", www.ieabioenergy.com/publications/bio-based-chemicals-value-added-products-from-biorefineries.

Anellotech of the United States has renewable chemistry solutions to the aromatic challenge. In its process, non-food biomass such as wood, sawdust, corn stover and sugar cane bagasse are gasified and immediately converted into hydrocarbons by a proprietary, reusable zeolite catalyst. The resulting mixture of benzene, toluene and xylenes (bio-BTX) is identical to the petroleum-derived counterparts.

The BTX compounds are integral to the production of a wide range of plastics including polyurethane, polycarbonate, polystyrene and nylon. Hence Toyota Tsusho and Anellotech have an alliance (Biofuels Digest, 2016): Toyota Tsusho is a multinational strategic equity investor in Anellotech and a corporate partner in the renewable aromatic chemicals supply chain. Aromatics are widely used in the automotive industry, and the Toyota Group has championed the use of renewables in vehicles (OECD, 2011).

This report frequently emphasises the alliance of industrial biotechnology with green chemistry. Their convergence has already solved challenges that one or the other could not solve alone. The aromatics challenge is another example of the need to support both, but it also reinforces the fact that solutions for biotechnology lag behind those for chemicals.

Bio-based production gaining visibility

For the public and policy makers, bio-based production has lacked visibility. Table 8.3, however, shows this visibility has increased dramatically in recent years. Nevertheless, this revolution in production could remain unheralded because a bio-based and fossil product look identical e.g. tyre, smart phone screen, drinks bottle. Certification and labelling would help improve this visibility enormously, giving confidence to manufacturers and helping with public perception and acceptance. The increased political impetus from 2015 onwards, especially COP21 and the drive towards a circular economy, could be used as levers to increase this visibility.

Brands and recent deals

The interest of brands has helped improve visibility as noted in Table 8.3. New business alliances ensure that new bio-based products are taking their place in the market (Box 8.1). Brands can also leverage their marketing and global outreach capacities to open markets for bio-based products.

Around 30 key bio-based chemicals are close to full market stability

European Commission (2015) reports more than 90 bio-based chemicals have reached a Technology Readiness Level (TRL) of at least 3. While there are only 3 such chemicals at TRL 9, there are 23 at TRL 8.5 and another 8 at TRL 8. According to EARTO classification (EARTO, 2014), this places them at least at the level of: "Manufacturing fully tested, validated and qualified", which agrees roughly with other TRL classification systems.[1] Therefore it would appear that a reasonable number of important bio-based chemicals are progressing towards TRL 9, which is effectively stable, competitive manufacturing. However, this says little about their market share or future prospects.

Many of these chemicals may not be recyclable or non-toxic; in many cases, they replace petro-based equivalents. The truly biodegradable, non-toxic ones usually take the same or similar function as a petro-based chemical. Their favourable GHG emissions compared to petro-counterparts is the overarching reason for their development.

A common denominator: The challenge of scale

Industry struggles to produce the vast majority of bio-based products and chemicals at a scale that can influence a market. For custom and specialty chemicals, the challenge is more easily surmounted than for commodity chemicals. Biofuels have proven difficult to transition from the laboratory to commercial production due to the huge volumes required to affect the market. In some countries, the margins on petrol and diesel production are so low that it remains difficult to make biofuels at a competitive price.

Table 8.3. **Bio-based products are becoming more familiar**

Latex from dandelions	Prototype tyres containing bio-based latex were showcased in December 2009 at the United Nations Climate Change Conference in Copenhagen. The Fraunhofer Society in partnership with the tyre company Continental has built a pilot plant to produce rubber from dandelions. The Russian dandelion thrives in soils unsuitable for agriculture.
Bottles from sugar	Both the Coca-Cola and PepsiCo companies have plastic bottles that are at least partly bio-based. The Coca-Cola bottle contains mono-ethylene glycol derived from fermented sugar. It is mixed with other components to make bio-polyethylene terephthalate. The long-term aim is to replace petro-PET. Avantium (Netherlands) and BASF intend to produce a different bioplastic for bottles (polyethylene furanoate).
Straw to fuel	In many OECD countries, bioethanol is moving from first generation (cellulosic feedstocks) to second generation. The first of the second-generation biorefineries are open. Clariant of Switzerland uses technology that breaks down lignocellulose enzymatically, and yeast ferments the sugars to ethanol.
Soybean to graphene	Graphene is more than 200 times stronger than steel and conducts electricity better than copper. About 1% of graphene mixed into plastics could turn them into electrical conductors. Graphene is, however, expensive compared to other materials. Researchers at CSIRO, Australia, have created a new method of graphene synthesis from soybean oil (Seo et al., 2017).
Castor nuts to wall plugs	DuPont extracts a chemical building block from castor oil to make a 68% bio-based polyamide, which is as strong as the nylon normally used to make wall plugs.
Bioplastics in cars	One of the earliest uses of bioplastics was replacing metal or petro-plastics components in vehicles, saving GHG emissions and/or weight. Among others, Ford and Toyota are investigating and using bioplastics as textiles in car interiors. Daimler and DSM worked together to create an engine cover that is a 70% bio-based plastic.
Sugar to carpets	Dupont and Mohawk combine bio-based propanediol and a petrochemical building block to make a carpet fibre that is soft, durable and easy to clean. The textile is 37% bio-based.
Yeast to face creams	Korres grows yeast cultures that produce hexapeptides when treated with ozone or irradiated with UV light. The compounds are added as anti-ageing active ingredients in face creams. Amyris has engineered specialised yeast strains that can produce squalene from sugar. Squalene is used as an emollient in moisturiser lotions (Servick, 2015).
Ice cream from lupins	Prolupin has developed a process to extract protein from the seeds of lupins. The protein is used to make ice cream that contains neither lactose nor gluten. Evolva uses a synthetic, biology-derived yeast for fermentation to synthetic vanillin. Other food materials through synthetic biology include stevia (sweetener) and nootkatone (smell of grapefruit).
Biopharmaceuticals	Antibiotics have been traditionally produced from microbes. Synthetic biology has been used to make a potent anti-malarial. Sanofi delivered the first large-scale batches of anti-malarial treatments manufactured with a new semi-synthetic artemisinin derivative to malaria endemic countries in Africa in 2014.
Bacteria in toothpaste	The probiotic Lactobacillus Pro-action, which can be added to toothpaste, specifically targets bacteria in the mouth that cause cavities. It can be added to toothpaste. The bacteria are produced by BASF and the toothpaste marketed by Neva Cosmetics.
Nutrition and food/feed supplements	Cargill makes a sweetener with a synthetic biology yeast to convert sugar molecules to mimic the properties of stevia, with no need for the plant itself. It awaits a commercial launch date. Calysta specialises in the production of microbial proteins for the commercial fish feed and livestock markets.
Enzymes in detergents	Biological detergents contain a range of enzymes that allow washing at lower temperatures, such as 30°C, thus saving energy, emissions and money.
Spider silk to medical implants	Spider silk, an exceptionally strong material, is used in sutures, scaffolds, grafts and some medical implants. Oxford Biomaterials, Orthox Ltd and Neorotex Ltd are investigating a range of biomedical applications of genetically engineered spider silk. The US army is testing protective garments for soldiers made from spider silk. An *E. coli* variant of spider silk could replace Kevlar in air bags.

Note: The first six examples are truly about replacements for petrochemicals, while the others demonstrate the eclectic range of bio-based possibilities. The source of this table gives more examples.

Source: Global Bioeconomy Summit (2015), "Bioeconomy in everyday life".

Box 8.1. Some recent business developments and alliances in bio-based production

February 2016. BRAIN Biotechnology Research and Information Network AG (BRAIN AG) had a stock market launch to become Germany's first listed bioeconomy company. Large parts of the chemical industry, in particular, have growth potential; experts foresee a rising share of biotechnology products and procedures. BRAIN AG focuses on specialty chemicals and the consumer chemicals divisions. The company received gross proceeds of EUR 31.5 million from the initial public offering (IPO). Deutsche Börse classified BRAIN as belonging to the speciality chemicals sector.

February 2016. Chinese renewable energy investment company Kaidi announced plans to build a biodiesel refinery in Finland. The value of the investment is EUR 1 billion, making it the biggest Chinese investment in Finland to date. The first of its kind, it will produce biofuels by using wood-based biomass. This includes energy wood, harvesting remains and even leftover bark from the forest industry as the main feedstock. The plant will produce 200 000 tonnes of biofuel per year, of which 75% will be renewable diesel and 25% renewable gasoline.

February 2016. Mitsui & Co., BioAmber's partner in the Sarnia (Canada) bio-based succinic acid plant, is investing an additional CAD 25 million in their joint venture. Mitsui will play a stronger role in the commercialisation of bio-succinic acid.

February 2016. Gevo, a renewable products and technology company, announced a license agreement and a joint development agreement with Porta Hnos, a leading alcohols company in Argentina, to construct multiple isobutanol plants in Argentina using corn as a feedstock.

March 2016. Air New Zealand and Virgin Australia announced a partnership to investigate options for locally produced aviation biofuel. The alliance partners are issuing a Request for Information (RFI) to the market to explore the opportunity to procure locally produced aviation biofuel.

April 2016. A new version of the Tetra Pak (Sweden) Tetra Top package will make its global debut in the United States. The new generation carton bottle now comes with a cap and top made from high-density polyethylene (HDPE) derived from sugarcane. Combined with the FSC-certified paperboard used in the main sleeve of the carton, this pushes its renewable content up from 53% to 82%, with no impact on its recyclability.

May 2016. Virent of Wisconsin, United States announced the world's first 100% plant-based polyester shirts. The development of the Virent technology platform is supported through strategic partners including Cargill, the Coca-Cola Company, Honda, Shell and Tesoro.

May 2016. Aemetis and Edeniq, both headquartered in California, entered into a definitive agreement under which Aemetis will acquire all of Edeniq's outstanding shares in a stock plus cash merger transaction. Aemetis is an advanced fuels and renewable chemicals company. Edeniq is a cellulosic ethanol technology company that has developed innovations that unlock cellulosic and starch sugars through a combination of mechanical and biological processes.

June 2016. PTT (formerly known as Petroleum Authority of Thailand) group joined with Japan's Mitsubishi Chemical Holding Corp to form a USD 100 million joint venture to build Thailand's first polybutylene succinate plant with an annual capacity of 20 000 tonnes.

July 2016. Ginkgo Bioworks and Amyris partnered to enable the companies to jointly develop products more efficiently and cost effectively, accelerating time to market. The deal aims to generate USD 300 million in incremental value. Ginkgo is building Bioworks2, a next-generation automated foundry where its organism engineers can develop new designs at massive scale. Amyris has commercialised five products from highly engineered organisms, going into markets from skin care and fragrances to industrial lubricants, tyres and jet fuel.

July 2016. The Ford Motor Company and Jose Cuervo announced an alliance to explore the use of the tequila producer's agave plant by-product to develop more sustainable bioplastics to employ in Ford vehicles.

August 2016. Amyris, in co-operation with Renmatix and Total New Energies in the United States, will work to develop a manufacturing-ready process using wood as the cellulosic feedstock to produce farnesene in a multi-million contract with the US DOE.

August 2016. Sacramento County, California, partnered with Neste of Finland for the trial supply of Neste renewable diesel in its fleet of more than 400 trucks and heavy equipment.

Box 8.1. **Some recent business developments and alliances in bio-based production** (*continued*)

September 2016. Toyobo, one of Japan's top fibres and textile manufacturers, and Avantium, a scale-up renewable chemicals company of the Netherlands, partnered on polyethylene furanoate (PEF) polymerisation and PEF films. The two companies have jointly developed thin films made from PEF, a 100% bio-based plastic. Avantium is working in collaboration with brand partners Danone and the Coca-Cola Company to bring 100% bio-based PEF bottles to the market.

September 2016. Neste of Finland and IKEA of Sweden announced a partnership to deliver renewable, bio-based plastics. The partnership combines IKEA's commitment to reduce dependence on virgin fossil-based materials and Neste's expertise in renewable solutions.

September 2016. LanzaTech has produced 1 500 gallons of jet fuel, derived from waste industrial gases from steel mills, via a fermentation process. The fuel has passed all its initial performance tests. It is the result of a partnership between Virgin and LanzaTech.

September 2016. Virent established a strategic consortium with Tesoro, Toray, Johnson Matthey and the Coca-Cola Company focused on completing the development and scale-up of Virent's BioForming technology to produce low-carbon bio-based fuels and bio-paraxylene (a key raw material for the production of 100% bio-polyester).

September 2016. Global Bioenergies, Preem, Sekab and Sveaskog announced having joined forces to develop a high-performance fuel entirely based on forest resources. The consortium has signed a collaboration agreement to carry out a conceptual scope study for a first plant in Sweden. This work will be carried out as part of the "Bio-Based Gasoline Project" with support from the Swedish Energy Agency.

September 2016. Mater Biotech, a 100% company owned by Novamont, opened its first commercial bio-BDO plant using Genomatica's technology that converts renewable feedstocks into 1,4 butendiol (BDO) in Bottrighe di Adria (Rovigo, Italy). Thanks to an investment of EUR 100 million, Novamont has managed to revive an abandoned manufacturing site of Bioitalia. The plant will produce 30 000 tonnes of renewable BDO per year by 2017.

September 2016. Loblaw of Canada announced the launch of compostable President's Choice (Loblaw's in-house brand) single-serve coffee pods. They are made almost entirely from plant materials and reclaimed coffee bean skins. They are the result of Canadian innovation and collaboration between the University of Guelph's Bioproducts Discovery and Development Centre (BDDC), Club Coffee (a Toronto-based company) and Competitive Green Technologies (Leamington, Ontario, a producer of bio-polymers/plastics and bio-composites).

October 2016. Ginkgo Bioworks and Genomatica announced an alliance to deliver biology-based solutions for the world's highest-volume intermediate and specialty chemicals more rapidly. Mainstream chemical producers can now in-license technology to manufacture their widely used chemicals with cost-effective and sustainable whole-process solutions that include engineered microorganisms, complete process designs and technology transfer support.

November 2016. The Danish Minister for Environment and Food launched the white paper on Danish circular economy at the conference "Danish Pioneers of Sustainability" hosted by the Confederation of Danish Industry.

November 2016. Global Bioenergies of France announced completion of its demonstrator plant in Leuna, Germany. This is the only facility in the world dedicated to the direct fermentation of gaseous hydrocarbons.

November 2016. Corbion of the Netherlands is building its new polylactic acid (PLA) bioplastics polymerisation plant at an existing Corbion site in Rayong, Thailand. Upon completion in 2018, it will be able to produce a portfolio of PLA neat resins: from standard PLA to innovative, high heat-resistant PLA.

December 2016. Leaf Resources of Australia announced a collaboration with Novozymes to further increase the yields and efficiency associated with Leaf Resources' innovative biomass conversion technology, Glycell, which is a combination of well-established process engineering and innovative chemistry.

December 2016. The South African Department of Agriculture, Forestry and Fisheries has approved four NexSteppe sorghum hybrids for commercial sale in the country. NexSteppe is a US company pioneering the next generation of sustainable feedstock solutions for the biofuels, biopower, biogas and bio-based products industries.

Box 8.1. **Some recent business developments and alliances in bio-based production** (*continued*)

January 2017. The US Department of Energy's Bioenergy Technologies Office (BETO) selected LanzaTech of New Zealand and the United States to receive USD 4 million to design and plan a demonstration-scale facility. They will use industrial off gases to produce 3 million gallons per year of low-carbon jet and diesel fuels. The facility will recycle industrial waste gases from steel manufacturing.

January 2017. In conjunction with the Institute for Materials and Wood Technology at the Bern University of Applied Sciences, AVALON Industries is launching a research project to replace formaldehyde in PF resins with the bio-based, non-toxic platform chemical 5-HMF (5-Hydroxymethylfurfural). Government-sponsored by the Swiss Commission for Technology and Innovation, the project will build on the positive results in a similar research project to develop non-toxic urea-HMF resins.

February 2017. Clariant of Switzerland, together with Mercedes-Benz and Haltermann Carless, tested the use of sustainable cellulosic ethanol from agricultural residues in a fleet test with Mercedes-Benz series vehicles over 12 months for the first time in Germany. The fuel by Haltermann Carless, which has a cellulosic ethanol content of 20% by volume (E20), was produced at Clariant's Sunliquid plant in Straubing, Germany. The cellulosic ethanol allows GHG emission savings of up to 95% across the entire value chain without competing with food production or tying up agricultural land.

February 2017. Global Bioenergies, France, announced the production of ETBE (ethyl-tertiary-butyl ether) purely from renewable resources. It can be used as an additive in vehicle fuel, up to a maximum of 23%, thereby increasing the proportion of biofuels in blends with fossil fuels. It is made by combining renewable ethanol with renewable isobutene. This first production of entirely renewable ETBE was supported by a grant of the German Ministry of Education and Research.

March 2017. Danone and Nestlé Waters, the world's two largest bottled water companies, have joined forces with Origin Materials, a Californian start-up, to form the NaturALL Bottle Alliance. Together, the three partners aim to develop and launch at commercial scale a 100% bio-based PET plastic bottle.

March 2017. The initial public offering (IPO) of Avantium raised EUR 103 million on Euronext Amsterdam and Euronext Bruxelles. Funds raised will be used to further commercialise Avantium's inventions into viable production processes. This will start with the commercialisation of the YXY technology, in a joint venture with BASF, by building the first commercial-scale reference plant for FDCA. On the basis of the share price, Avantium's market capitalisation reached EUR 277 million.

On the other hand, high-value specialty and fine chemicals are mostly produced in more manageable, low volumes (and market sizes) with which a young industry can cope. They also offer larger margins. The successful production of low-volume chemicals via metabolic engineering routes may provide greater market confidence than failure to make high-volume fuels. Companies adopting this strategy may be considered as the second generation of synthetic biology, or metabolic engineering, companies.

Even if successful in the marketplace, high-value speciality and fine chemicals may not have a huge impact on overall GHG emissions. Large volume, low margin commodity chemicals generally generate the largest GHG emissions. In the analysis by Saygin et al. (2014), seven bio-based materials had an estimated technical CO_2 emissions reduction potential of 0.3-0.7 Gigatonnes (Gt) CO_2 in 2030. Assuming the same potential for the remainder of organic materials production, they estimated a total technical reduction potential of up to 1.3-1.4 Gt CO_2 per year by 2030 compared to 3.2-3.7 Gt CO_2 for fuels.

Conclusions

The nascent bio-based materials industry has accomplished much of its achievements with little policy support beyond a subsidy for research and development (OECD, 2014). This is understandable as there is a mere handful of liquid fuels and vast numbers of chemicals, complicating the possibilities for mandates. However, not supporting bio-based materials in public policy misses significant opportunities for GHG savings. It also fails to take advantage of other policy goal benefits such as making a good fit with circular economy ambitions, reindustrialisation and decentralised manufacturing. These policy goals find excellent alignment with the integrated biorefinery concept, the most ambitious, but also most complex, biorefinery model. Ignoring bio-based chemicals and materials in public policy makes the economics of integrated biorefineries questionable; the margins for many chemicals are usually better than for high-volume fuels. The widespread policy support for biofuels and bioenergy systematically allocate biomass for these purposes, and not for materials.

Note

1. Technology Readiness Levels (TRLs) are a method of estimating technology maturity, generally ranging from 1 (basic research) to 9 (launch and operations): https://en.wikipedia.org/wiki/Technology_readiness_level.

References

Aeschelmann, F. et al. (2015), *Bio-based Building Blocks and Polymers in the World – Capacities, Production and Applications: Status Quo and Trends towards 2020*, Nova-Institut GmbH, Chemiepark Knapsack, Köln, Germany.

ANZ Insights (2012), "Global plastics industry: Market update", ANZ Research.

AT Kearney (2012), *Chemical Industry Vision 2030: A European Perspective*, AT Kearney, https://www.atkearney.com/chemicals/article?/a/chemical-industry-vision-2030-a-european-perspective (accessed 15 January 2018).

Biofuels Digest (2016), "Toyota: Let's go places with Anellotech, bio-places that is", *Biofuels Digest*, 17 July.

Broeren, M.L.M. et al. (2014), "Forecasting global developments in the basic chemical industry for environmental policy analysis", *Energy Policy*, Vol. 64, Elsevier, Amsterdam, pp. 273-287.

Delucchi, M.A., et al. (2014), "An assessment of electric vehicles: Technology, infrastructure requirements, greenhouse gas emissions, petroleum use, material use, lifetime cost, consumer acceptance and policy initiatives", *Philosophical Transactions of the Royal Society A*, 2 December, Royal Society Publishing, London, pp. 372: 20120325, http://dx.doi.org/10.1098/rsta.2012.0325.

EARTO (2014), "The TRL scale as a research & innovation policy tool, EARTO recommendations", 30 April, European Association of Research and Technology Organisations, Brussels.

Eriksson, N. (2013), "Production of four selected renewable aromatic chemicals", Chalmers University of Technology, Gothenburg, Sweden.

European Commission (2015), "From the sugar platform to biofuels and biochemical", Final report for the European Commission Directorate-General Energy, ENER/C2/423-2012/SI2.673791, April, European Commission, Brussels.

Frost, J.W. and J. Lievense (1994), "Prospects for biocatalytic synthesis of aromatics in the 21st century", *New Journal of Chemistry*, Vol. 18/3, Royal Society of Chemistry, London, pp. 341-348.

Global Bioeconomy Summit (2015), "Bioeconomy in everyday life", German Bioeconomy Council, Berlin.

Hellsmark, H. et al. (2016), "Innovation system strengths and weaknesses in progressing sustainable technology: The case of Swedish biorefinery development", *Journal of Cleaner Production*, Vol. 131, Elsevier, Amsterdam, pp. 702-715.

Hopewell J. et al. (2009), "Plastics recycling: Challenges and opportunities", *Philosophical Transactions of the Royal Society B*, 364, Royal Society Publishing, London, pp. 2115-2126.

IEA (2012), *Energy Technology Perspectives 2012 – Pathways to a Clean Energy System*, International Energy Agency, Paris, http://dx.doi.org/10.1787/energy_tech-2012-en.

IEA Bioenergy Task 42 Biorefinery (2012), "Bio-based chemicals. Value added products from biorefineries", International Energy Agency, Paris, www.ieabioenergy.com/publications/bio-based-chemicals-value-added-products-from-biorefineries.

Il Bioeconomista (13 September 2016), "KLM signs a three-year contract for the supply of biofuel in Los Angeles", Il Bioeconomista blog, https://ilbioeconomista.com/2016/09/13/klm-signs-a-three-year-contract-for-the-supply-of-biofuel-in-los-angeles.

IRENA/ETSAP (2013), "Production of bio-ethylene. Technology Brief", www.irena.org/DocumentDownloads/Publications/IRENA-ETSAP%20Tech%20Brief%20I13%20Production_of_Bio-ethylene.pdf.

Jiménez-Sánchez, G. and J.C. Philp, (2015), "Omics and the bioeconomy: Applications of genomics hold great potential for a future bio-based economy and sustainable development", *EMBO Reports*, Vol. 16/1, European Molecular Biology Organization, Heidelberg, pp. 17-20.

Kawaguchi, H. et al. (2016), "Bioprocessing of bio-based chemicals produced from lignocellulosic feedstocks", *Current Opinion in Biotechnology*, Vol. 42, Elsevier, Amsterdam, pp. 30-39.

Milken Institute (2013), "Unleashing the power of the bio-economy", February, Santa Monica, United States.

Moreau, N.J. (2005), "Public images of chemistry", *Chemistry International*, Vol. 27/4, July-August, International Union of Pure and Applied Chemistry, Research Triangle Park, United States.

OECD (2014), "Biobased chemicals and plastics. Finding the right policy balance", *OECD Science, Technology and Industry Policy Papers*, No. 17, OECD Publishing, Paris, http://dx.doi.org/10.1787/5jxwwfjx0djf-en.

OECD (2013), "Policies for bioplastics in the context of a bioeconomy", *OECD Science, Technology and Industry Policy Papers*, Vol. 10, OECD Publishing, Paris, http://dx.doi.org/10.1787/5k3xpf9rrw6d-en.

OECD (2011), *Future Prospects for Industrial Biotechnology*, OECD Publishing, Paris, http://dx.doi.org/10.1787/9789264126633-en.

Philp, J.C. (2015), "Balancing the bioeconomy: Supporting biofuels and bio-based materials in public policy", *Energy & Environmental Science*, Vol. 8, Royal Society of Chemistry, London, pp. 3063-3068.

Saygin, D. et al. (2014), "Assessment of the technical and economic potentials of biomass use for the production of steam, chemicals and polymers", *Renewable and Sustainable Energy Reviews*, Vol. 40, Elsevier, Amsterdam, pp. 1153-1167.

Scania (n.d.), "Take control: Hybrid truck for city use", webpage, www.scania.com/group/en/take-control-hybrid-truck-for-city-use (accessed 7 December 2017).

Seo, D.H. et al. (2017), "Single-step ambient-air synthesis of graphene from renewable precursors as electrochemical genosensor", *Nature Communications*, Vol. 8, Springer Nature, pp. 14217.

Servick, K. (15 October 2015), "Four synthetic biology inventions that flummox the feds", Science News blog, American Association for the Advancement of Science, Washington, DC, http://dx.doi.org/10.1126/science.aad4723.

Smolarski, N. (2012), "High-value opportunities for lignin: Unlocking its potential", Frost & Sullivan, www.greenmaterials.fr/wp-content/uploads/2013/01/High-value-Opportunities-for-Lignin-Unlocking-its-Potential-Market-Insights.pdf (accessed 31 January 2018).

SkyNRG (n.d), SkyNRG website, http://skynrg.com (accessed 7 December 2017).

Takors, R. (2012), "Scale-up of microbial processes: Impacts, tools and open questions", *Journal of Biotechnology*, Vol. 160/1-2, Elsevier, Amsterdam, pp. 3-9.

USDA (1956), "Hardwood-distillation industry", No. 738, United States Department of Agriculture, Forest Products Laboratory, Madison, United States.

US DOE (2004), "Top value-added chemicals from biomass (results of screening for potential candidates from sugars and synthesis gas, Vol. 1)", Werpy, T. and G. Petersen (eds.), United States Department of Energy, Washington, DC.

Wong, K.-C. (ed.) (2016), *Computational Biology and Bioinformatics: Gene Regulation*, CRC Press, Taylor & Francis Group.

Yim, H. et al. (2011), "Metabolic engineering of Escherichia coli for direct production of 1,4-butanediol", *Nature Chemical Biology*, Vol. 7, Nature Publishing Group, London, pp. 445-452.

Chapter 9.

Enabling bio-based materials policy

In recent years, the absence of policy support for bio-based chemicals and materials production in the face of huge support for both biofuels and bioenergy has been a matter of contention. This lopsided emphasis has serious consequences for integrated biorefineries of the future. It systematically allocates (subsidised) biomass to fuels and energy applications; as a result, opportunities for high value-added and greater job creation could be missed. If lessons from petro-fining are any indication, lack of support for bio-based chemicals and materials production may completely throw the economics of integrated biorefinery operation into doubt. This chapter examines policy options that will start to address the situation from economic, environmental and social perspectives. It aims to help governments implement policy support for bio-based materials that can be consistent with that for national biofuels. This would be a cost-efficient mechanism that uses existing support policies and conditions rather than creating a separate support scheme with its own infrastructure and bureaucracy.

Introduction: Improving bio-based materials policy

For several years, many publications (e.g. Snyder, 2015) and events (e.g. Friends of Europe, 2012) have argued for a "level playing field" for bio-based materials (mainly bio-based plastics and chemicals). This argument refers to the large and widespread support given to biofuels and bioenergy in many countries as part of their obligations to reduce emissions of greenhouse gases (GHGs). In most countries with biofuels and bioenergy policies, public policy support for bio-based materials has been all but absent. Support that has been given has often been limited to research and development (R&D) subsidy.

Greater attention for bio-based materials is important, in large part, to make the integrated biorefineries of the future economically viable. Much of the profit would come from the lower production volumes of chemicals because their margins are generally superior to those of fuels. Not supporting bio-based materials in policy runs the risk that integrated biorefineries will not be able to function profitably.

The starting point is the US Renewable Fuel Standard (RFS). This mandates biofuels production targets through to 2022, but also sets GHG emissions targets for each category of biofuel included in the mandate (see Box 9.1). To guarantee improved environmental performance, the RFS mandates steadily increasing production of biofuels with superior GHG emissions reductions, especially cellulosic ethanol. At the same time, it allows corn-based bioethanol (first-generation bioethanol) to reach a plateau (Figure 9.1).

Figure 9.1. **Renewable Fuel Standard mandated production through to 2022**

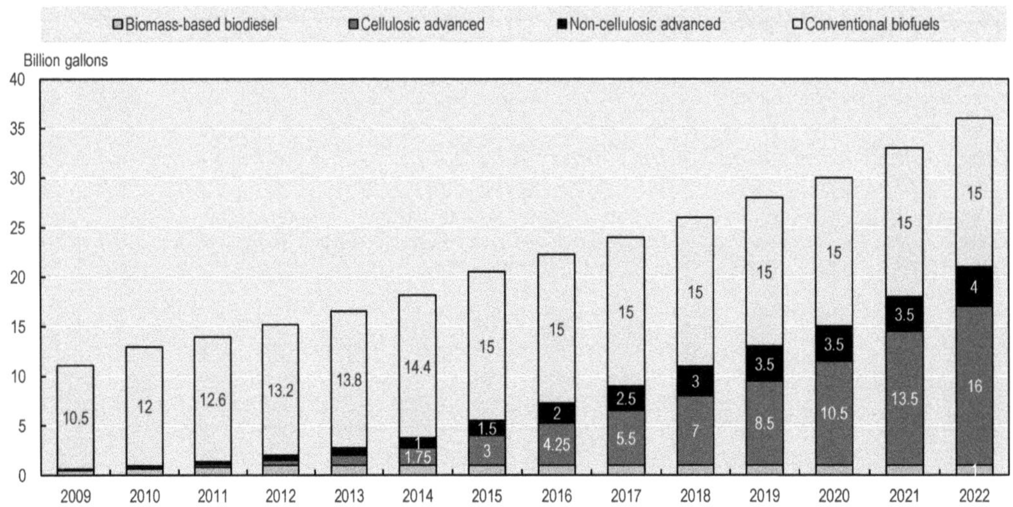

Note: There have been notable setbacks in US biofuels production among some of the ethanol categories that have delayed policy decisions. This is particularly true of cellulosic advanced ethanol.

Source: Redrawn from US EPA (2010), "Renewable Fuel Standard Program", https://www.epa.gov/renewable-fuel-standard-program/overview-renewable-fuel-standard.

Bio-based materials have a similar policy goal – to support the development of materials with better environmental performance. The greatest reductions in emissions would be gained from the bio-based equivalents of large production volume commodity chemicals. Therefore, a policy similar to RFS for materials would provide similar benefits, albeit that the scales of production are far lower than for fuels.

Box 9.1. The Renewable Fuel Standard and mandated targets for biofuels production

The Energy Independence and Security Act (EISA) set minimum volumes of renewable fuels that suppliers must blend into the US supply of transportation fuel each year, irrespective of market prices. This effectively guarantees a market for biofuels. The Renewable Fuel Standard 2 (RFS2) substantially reduces the risk associated with biofuels production. In so doing, it provides an indirect subsidy for capital investment in the construction of biofuels plants. As such, the expanding RFS is expected to continue to stimulate growth of the biofuels industry.

EISA requires that emissions associated with a renewable fuel are at least a certain percentage lower than those associated with the gasoline or diesel that it replaces (US EPA, 2009). EISA therefore attempts to address energy security, rural regeneration and climate change mitigation, while growing a large number of jobs in the ethanol industry.

The Environmental Protection Agency (EPA) establishes and implements regulations to ensure that the nation's transportation fuel supply contains the mandated biofuels volumes (CRS, 2013). The EPA translates the yearly volume requirements in EISA into percentage standards (sometimes called blend requirements). These are based on projections of the total amount of gasoline and diesel that will be used in that year. For example, if the projected amount was 100 billion gallons and the total renewable fuel requirement was 14 billion gallons, the EPA would set a 14% blend requirement (CBO, 2014).

To monitor suppliers' compliance with the requirements, the EPA assigns a unique "renewable identification number" (RIN) to each qualifying gallon of renewable fuel. Every RIN includes a code that identifies which of the four RFS categories – total renewable fuels, advanced biofuels, cellulosic biofuels or biomass-based diesel – the gallon satisfies. Each fuel supplier, regardless of what kind of fuel it produces or imports, must meet all of the blend requirements for a given compliance year.

The supplier achieves compliance by using the required amounts of renewable fuels itself and submitting the corresponding RINs to EPA, by purchasing RINs from other suppliers that have excess RINs to sell or by submitting RINs that it acquired in the previous year and saved for future use. For the example above, each fuel supplier would have to submit 14 RINs (including 4 for advanced biofuels and 2 for biomass-based diesel) for each 100 gallons of gasoline or diesel that it sold. Suppliers with excess biomass-based diesel RINs could either sell them or apply them towards their advanced biofuel requirement.

The huge variety of chemicals that exists compared to fuels – some 70 000 products – makes it difficult to establish a single policy for bio-based chemicals. On the one hand, making ethanol from yeast is a relatively efficient bioprocess; yeasts can achieve high concentrations of ethanol in solution, and ethanol downstream purification is tried and tested. For many other bio-based chemicals, however, this is certainly not the case. The cascading policy options outlined here attempt to address both these issues and GHG emissions.

Policy design

The policy suggestions here essentially combine elements of industrial and green growth policy. The issue is about creating new manufacturing opportunities that allow economic growth, while avoiding the trap of increased emissions (UNEP, 2010). This was at the heart of the 2009 OECD publication *The Bioeconomy to 2030: Designing a Policy Agenda.*

General points

Good policy design should ensure competitive selection processes; contain costs; and select projects that best serve public policy objectives, without favouring incumbents or providing opportunities for lobbying (OECD, 2013). This suggests the need for a portfolio of public investment where funding approaches are tailored to the different stages of technology development. The technology development spans virtually the whole range of 1-9 of Technology Readiness Levels (TRLs) as each is designed on a one-off basis. Therefore, this point for policy makers is especially pertinent – policy for bio-based materials must be flexible enough to cover a wide range of technology readiness.

In general, policies for innovation and deployment need to encourage experimentation. These experiments should develop new options that can help strengthen environmental performance at the lowest cost (OECD, 2013). Given the early stage development of bio-based materials, policies need to trigger the industry to innovate continuously. Ultimately, it needs to develop improved bio-based alternatives to achieve ambitious CO_2 emissions reductions (Saygin et al., 2014).

Governments should level the playing field between alternative options. In general, however, it should avoid championing specific technologies and solutions, emphasising competition and technology neutrality. Other sources of organic chemicals in future manufacturing should not be excluded in favour of bio-based. Nevertheless, the sources of carbon for sustainably produced organic chemicals seem limited. Petrochemical manufacturing will continue to be important, but it is ultimately unsustainable. The only foreseeable alternative sustainable source to bio-based is waste CO_2 itself, as part of the CO_2 economy (GreenFire Energy, n.d.).

Against a background that no single technology or policy will drive green innovation, Dutz and Pilat (2014) recommended that countries combine supply- and demand-side policy instruments to achieve policy goals, which may differ from country to country. This is consistent with the conclusion by Mowery and Rosenberg (1979) that instruments related to both supply and demand are necessary for innovation. The OECD publication *Demand-side Innovation Policies* (OECD, 2011) details the relationship between supply- and demand-side policy to stimulate innovation.

How to tackle thousands of different chemicals

Thousands of different chemicals are manufactured from oil. Even the list of "significant" chemicals (in terms of production volume) runs to dozens (Wikipedia, n.d.). Creating a policy support mechanism akin to the feed-in tariff used successfully for renewable electricity is nigh on impossible for chemicals. Also, there is a mere handful of large-volume liquid fuel types, which greatly simplifies creating production mandates for biofuels. Attempting production mandates for individual chemicals would most likely meet with industry resistance due to the bureaucratic burden and cost.

Carus et al. (2014) suggest an innovative mechanism that would avoid creating and administering individual mandates or quotas for large numbers of different chemicals: using bioethanol as a reference chemical. Ethanol made using certified sustainable biomass, then used to manufacture chemicals and plastics, could be counted in the same way that ethanol is counted for a biofuel. All other bio-based chemicals not derived from ethanol, such as lactic acid, could be converted to ethanol "equivalents" (e.g. calorie value or molecular weight or number of carbon atoms compared to ethanol). This simple algorithm avoids dealing with many chemicals individually.

Such an approach would imply that chemicals "larger" than ethanol (i.e. a higher calorie value; larger number of carbon atoms amount to the same thing) would have a greater subsidy. While larger number of carbon atoms may mean greater sequestration of carbon in a chemical, this is not necessarily so: not all bio-based materials are synthesised entirely of bio-based carbon. Therefore, more detailed environmental performance data for the chemical are needed for policy making. Harmonised life cycle analysis (LCA) procedures are needed to calculate the emissions savings, which would become the basis for policy support. If the molecule in question is only partly bio-based, the percentage should be made clear – this could provide the stimulus for improved bio-based content.

Setting target environmental performance threshold levels

The Renewable Fuel Standard set GHG emissions reduction thresholds for different categories of biofuels (Table 9.1). This provides the stimulus for improvements in environmental performance. Thresholds could be set for bio-based materials in a similar manner so that:

- Public R&D funds, and potentially public contributions to scale-up (through, for example, loan guarantees and other PPP mechanisms), are directed to improving environmental performance.
- Projects are selected based on combined merits of environmental and economic attributes.
- Producers are encouraged to continuously strive for improvements through funding R&D.

Table 9.1. **GHG emissions reduction values specified for the Renewable Fuel Standard**

Fuel	GHG threshold (EISA)[1]
Renewable fuel	20%
Advanced biofuel	50%
Biomass-based diesel	50%
Cellulosic biofuel	60%

1. Percentage reduction from 2005 baseline.

Note: GHG = greenhouse gas; EISA = Energy Independence and Security Act.

Source: US EPA (2009), "EPA proposes new regulations for the national Renewable Fuel Standard program for 2010 and beyond".

However, Weiss et al. (2012) point out that large degrees of error in assessment of the GHG savings for bio-based materials is a major barrier to setting thresholds. LCA has created inconsistencies in approach, and its shortcomings have been summarised recently (OECD, 2014).

Saygin et al. (2014) selected the seven most important bio-based materials that could technically replace half of petrochemical polymers and fibre consumption worldwide. With these materials, they estimated a technical CO_2 emissions reduction potential of 0.3-0.7 Gigatonnes (Gt) CO_2 in 2030. Assuming the same potential for the remainder of organic materials production, they estimated a total technical reduction potential of up to 1.3-1.4 Gt CO_2 per year by 2030. With process improvements, they estimate 1.7-1.9 Gt CO_2 per year. These figures are compared to the emissions savings from fuel in Table 9.2.

Table 9.2. **Technical and economic potentials for CO_2 emissions reductions in 2030 and 2050**

Gt CO_2 per year

Biomass use	Technical potential (with autonomous improvements)		Economic potential (with energy efficiency)	
	2030	2050	2030	2050
Feedstock	1.3-1.4	1.7-1.9	0.3-0.4	0.6-0.7
Fuel	3.2-3.7	3.4-4.1	0.8-1.0	1.3-1.6
Total	4.5-5.1	5.1-6.0	1.2-1.3	1.9-2.2

Note: These potentials exclude biomass use in the pulp and paper sector.

Source: Extracted from Saygin et al. (2014), "Assessment of the technical and economic potentials of biomass use for the production of steam, chemicals and polymers".

Overall, in terms of generating steam, they conclude that some bio-based materials score better than biomass, while others score worse. Therefore, in the near future, policies have to reflect this variability, recognising that biomass supply will be limited. Decisions can be made based on efficiency of use and best use of public money, and provide guidance to business and consumers.

Table 9.1 suggests threshold levels of RFS for a first draft of a tool that could help governments make specific decisions. Further research would elucidate if these are appropriate levels. In the immediate term, these levels would allow seamless entry of bio-based materials into biofuels policy. Such a policy should be kept flexible to take account of future innovations to prevent inappropriate lock-in. In other words, future developments are likely to drive improved GHG emissions reductions. Policy should allow for change in threshold values to drive these improvements.

Taking account of production volume

The production volume of a chemical becomes relevant when considering its environmental impact through total emissions savings: as production volume increases so do potential savings. LCA may determine that a chemical has great potential for GHG savings. However, if it is a high-value chemical of low production volume, it has a limited overall contribution in terms of tonnes of CO_2 saved per year.

The nascent bio-based industry has come up against a serious barrier that creates a conundrum for policy making. Trying to make a high-volume, bio-based equivalent of a petrochemical suffers two large impediments.

First, over decades, the petrochemical equivalent has had its production process and supply chains perfected and the production plants have been amortised; as a result, it benefits enormously from economies of scale. A bio-based equivalent would find it difficult indeed to compete on price. It would be easier to compete on price with a low-volume, high-value chemical.

Second, bioprocesses are notoriously inefficient when it comes to scaling up to a level that can influence a market. Microorganisms have not evolved to work in the severe environment of a bioreactor. Hence, serious modification is virtually always required to achieve the titre and yield necessary to make it economical. This modification is an iterative process that can have long innovation cycles to achieve high efficiency: it took the industry giants DuPont and Genencor approximately 15 years and 575 person years to develop and produce 1,3-PDO (Hodgman and Jewett, 2012). It takes on average 7.4 years to launch a bio-based product (Il Bioeconomista, 2015).

Naturally, small companies trying to make a bio-based chemical commercially opt for high-value chemicals that have low enough production volume to influence the market. But there is a conundrum. For policy makers, replacing the oil barrel requires bio-based alternatives to the major petrochemicals such as ethylene and other short-chain olefins.

As a policy option, one stage in decision making could allow for total global production volume that triggers a threshold for policy support: lower support for lower production volume, greater support for higher volume. This makes sense in the policy setting as greater production volume means greater potential GHG emissions saving, therefore higher value in climate change mitigation. This option is particularly attractive for nations using the mechanism to help meet emissions targets; it should act as the sought-after R&D stimulus for companies to make process improvements. This, in turn, will lead eventually to large-volume bio-based equivalents becoming competitive at scale.

Such a strategy would, of course, differ in different countries. For some countries, a balanced portfolio of investments in high and low production volume products is already a high priority.

Production efficiency factors

By specifically increasing the titre (g per litre of product), yield (g product per g substrate, normally glucose) and productivity (g per litre per hour), manufacturers and policy makers obviously benefit. This is preferred to industry and policy being at loggerheads. Lower water and energy requirements are the major outcomes, which mean improved sustainability, with two-way benefits. Here are some examples why:

- Lower volumes of water to recycle and treat can mean lower CO_2 emissions, especially if biological wastewater treatment is involved.
- Lower energy requirements are needed for smaller bioreactors with less water as the final product is more concentrated at the end of fermentation.
- Less water must be pumped around, less energy is required for reactor heating and/or cooling.
- Less energy is required for cleaning in place and sterilisation in place.
- Down-time between batches would be lower, and maintenance turnaround quicker.
- Higher titre means the product is more concentrated, so the process requires less energy input for downstream processing (purification from a very dilute solution can be enormously expensive).

What is more, creating a factor that improves production efficiency in this manner stimulates the research that policy makers want – research leading to lower marginal production cost. And, rather than paying through a subsidy, R&D tax credits or production tax credits may be able to cover public cost, depending on eligibility. This, in the longer term, would be a more palatable mechanism than mandated production.

It is not enough to modify the hardware of the bioprocess to bring about improvement. Biocatalyst genetic engineering and synthetic biology are likely to take improvements much further than can be achieved with reactor design. For example, consolidated bioprocessing (CBP) refers to combining lignocellulosic conversion to fermentable sugars[1] within the same microorganism that converts the sugars to bio-based products. CBP technology is widely considered the ultimate low-cost configuration for cellulose hydrolysis and fermentation (US DOE, 2006).

Summary

A cascading policy support mechanism (Figure 9.2) would bring bio-based materials under the umbrella of biofuels support. Its construction addresses both environmental performance and cost-efficiency for the taxpayer. It could also stimulate R&D towards making the most efficient bio-based chemicals (in terms of GHG emissions reductions) in the most efficient bioprocess (in terms of cost for the manufacturer). It specifically addresses high-volume, low-value chemicals because these have the greatest impact in replacing the oil barrel and in emissions reduction. These are precisely the chemicals that do not attract the young bio-based industry due to the difficulty to synthesise them efficiently at scale in competition with the petrochemicals industry.

Figure 9.2. **A generic decision support cascade for embedding bio-based materials policy support within biofuels support**

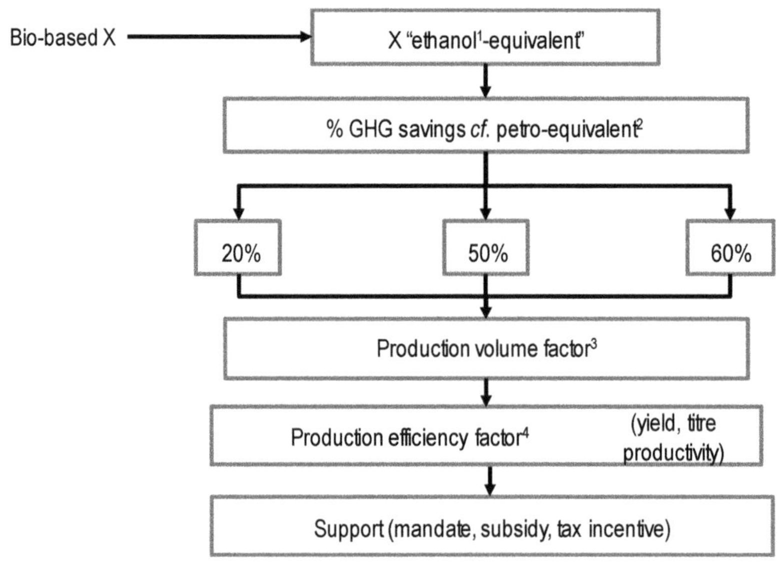

1. Consider also "lactic acid-equivalent".

2. EISA biofuels reference: renewable fuel = 20%, advanced biofuel = 50%, biomass-based diesel = 50%, cellulosic biofuel = 60%.

3. Small volumes will not have significant total GHG emissions savings, i.e. they are inefficient.

4. Encourages innovation to improve efficiency.

Note: GHG = greenhouse gas.

Strengths

- This rationalises the potentially many chemicals into a single equivalent that is the industry standard (bioethanol) and that already exhibits high bioprocess production efficiency.
- It includes two measures designed to improve environmental performance. The first, in this generic scheme, uses the same GHG emissions standards as in the model biofuel policy (RFS), but is adaptable to any national/regional standards. The second takes account of the potential global GHG savings for any particular chemical that can be easily derived using the global production capacity. Both measures allow flexibility in the event of changes to GHG emissions standards and/or global production tonnages.

- It should drive innovation to improve the efficiency of bioprocesses for large-volume, low-value chemicals, precisely the ones that are most difficult without policy.
- It should make best use of public money by removing replication of bureaucracy.
- It would avoid or minimise some significant issues around ethanol as a biofuel due to, among other things, the much smaller production volumes of chemicals compared to fuels. Examples are imagined or real food prices impacts;[2] blend wall is not an issue; limited impacts on transportation infrastructure (e.g. no need for new pipelines) and no fuel stations infrastructure issues; less complex demand-side issues (e.g. no flex-fuel vehicles).

Mitigating the weaknesses

- As it stands, the cascade does not include two important technology categories for renewable chemicals: those produced through waste CO_2, and those that can be produced either entirely by "green chemistry" or by a combination of bio-based and green chemical technologies. However, if the GHG emissions reductions for chemicals produced by these technologies are known, they should be rather easily incorporated into the scheme. The best example is bio-based ethylene, the synthesis of which involves fermentation to ethanol followed by chemical conversion to ethylene.
- It does not specify eligibility for entry to the scheme. However, it is intended for production rather than R&D, although eligibility for chemicals that need some near-market R&D is suggested, depending on state-aid rules. Therefore, it would seem sensible to make the scheme eligible to chemicals at a TRL of 7 and above in the US Department of Defense classification (US DOD, 2011). Or simply, the policy could specify technologies that are "beyond demonstration".
- The chemicals described are identical, drop-in replacements for petrochemicals, and therefore are not "needed" as such. In RFS, ethanol is desirable in petrol (gasoline) as a fuel oxygenate. Therefore, this would justify the petrochemicals industry accepting such a policy.
- Such a policy cannot be brought in for many bio-based chemicals as few are produced at volume. This is part of the reason for the policy – to stimulate greater production of a greater number of bio-based chemicals. Therefore, a phased approach would be needed. Each country would need to decide which chemicals to concentrate on, and slowly add to its inventory by keeping the policy flexible. This could be co-ordinated with a national bioeconomy strategy and/or a national biorefinery roadmap. However, it would be difficult to specify a date when the mandate ends or how the mandate may be phased out; it must do this to remove longer-term market distortion.
- The position of large polymers is not clear. Large bio-based equivalents of thermoplastics would sequester a lot of carbon, and policy may not reflect this. However, it would be reflected in the global production volume of the monomer. For example, ethylene is the largest production volume organic chemical. It is subsequently polymerised to polyethylene. If the manufacturers of ethylene and polyethylene are different, then one or the other may not qualify in this scheme. Both cannot qualify, as this would amount to double counting. This is because the polymerisation stage does not use any new bio-based carbon; it uses the bio-based carbon in the bio-ethylene.

Conclusions

Creating a level playing field between bio-based materials, biofuels and bioenergy has stayed as a defining topic in bioeconomy arguments. The potential solution laid out here in basic terms could address the need. Each country would need to develop the idea to suit its own conditions – after all, different countries have their own strengths and weaknesses. Making integrated biorefineries viable depends on balancing materials, fuels and energy production. Work by individual countries could mitigate any weaknesses. The scheme could be a cost-effective way forward as it simplifies bureaucracy and infrastructure for policy implementation.

Notes

1. One of the most significant challenges in using the vast global lignocellulose resource is the need for large quantities of enzymes to efficiently convert lignocellulose, hemicellulose and cellulose into fermentable sugars. These enzymes represent the second highest contribution to raw material cost after the feedstock itself.

2. There is a link between "imagined or real food price impacts" and food production. Specifically, one aspect surrounding corn-based ethanol is the diversion of grain for feed (primarily for cattle) to fuel. This avenue, and there could be others, can lead to price impacts.

References

Carus, M. et al. (2014), "Proposals for a reform of the Renewable Energy Directive to a Renewable Energy and Materials Directive (REMD). Going to the next level: Integration of bio-based chemicals and materials in the incentive scheme", *nova-Paper on bio-based economy,* No. 4, Nova-Institut, Huerth, Germany.

CBO (2014), "The Renewable Fuel Standard: Issues for 2014 and beyond", *CBO Publication*, No. 4765, Congressional Budget Office, Washington, DC.

CRS (2013), "Renewable Fuel Standard (RFS): Overview and issues", Report, No. R40155, Congressional Research Service, Washington, DC.

Dutz, M.A. and D. Pilat (2014), "Fostering innovation for green growth: Learning from policy experimentation", in *Making Innovation Policy Work*, OECD/World Bank, Paris, http://dx.doi.org/10.1787/9789264185739-10-en.

Friends of Europe (2012), "The rise of the bio-based economy", *Report of the Greening Europe Forum GEF) roundtable*, co-organised by Friends of Europe and Novozymes, Friends of Europe, Bibliothèque Solvay, Brussels.

GreenFire Energy (n.d.), "The new CO_2 economy," webpage, www.greenfireenergy.com/the-co2-economytrade.html (accessed 7 December 2017).

Hodgman, C.E. and M.C. Jewett (2012), "Cell-free synthetic biology: Thinking outside the cell", *Metabolic Engineering*, Vol. 14/3, Elsevier, Amsterdam, pp. 261-269.

Il Bioeconomista (10 June 2015), "Synbio start-ups need an average of 7.4 years to launch first chemical product", Il Bioeconomista blog, http://ilbioeconomista.com/2015/06/10/synbio-start-ups-need-an-average-of-7-4-years-to-launch-first-chemical-product.

Mowery, D. and N. Rosenberg (1979), "The influence of market demand upon innovation: A critical review of some empirical studies", *Research Policy*, Vol. 8, Elsevier, Amsterdam, pp. 102-153.

OECD (2014), "Sustainable biomass drives the next bioeconomy: A new industrial revolution?", *Report of an OECD Workshop*, Paris, 10-11 June 2014, OECD, Paris.

OECD (2013), "Beyond industrial policy: Emerging issues and new trends", *OECD Science, Technology and Industry Policy Papers*, No. 2, OECD Publishing, Paris, http://dx.doi.org/10.1787/5k4869clw0xp-en.

OECD (2011), *Demand-side Innovation Policies*, OECD Publishing, Paris, http://dx.doi.org/10.1787/9789264098886-en.

Saygin, D. et al. (2014), "Assessment of the technical and economic potentials of biomass use for the production of steam, chemicals and polymers", *Renewable and Sustainable Energy Reviews*, Vol. 40, Elsevier, Amsterdam, pp. 1153-1167.

Snyder, S.W. (2015), *Commercializing Biobased Products: Opportunities, Challenges, Benefits and Risks*, Royal Society of Chemistry, London.

UNEP (2010), *Assessing the Environmental Impacts of Consumption and Production: Priority Products and Materials*, United Nations Environment Programme, Washington, DC.

US DOD (2011), *Technology Readiness Assessment (TRA) Guidance*, United States Department of Defense, Washington, DC.

US DOE (2006), "Breaking the biological barriers to cellulosic ethanol: A joint research agenda", DOE/SC-0095, US Department of Energy Office of Science and Office of Energy Efficiency and Renewable Energy.

US EPA (2010), "Renewable Fuel Standard Program", webpage, https://www.epa.gov/renewable-fuel-standard-program/overview-renewable-fuel-standard (accessed 31 January 2018).

US EPA (2009), "EPA proposes new regulations for the national Renewable Fuel Standard program for 2010 and beyond", United States Environmental Protection Agency, Washington, DC.

Weiss M. et al. (2012), "A review of the environmental impacts of bio-based materials", *Journal of Industrial Ecology*, Vol. 16/1, Wiley Online Library, pp. S169-S181.

Wikipedia (n.d.), "Petrochemicals", webpage, http://en.wikipedia.org/wiki/Petrochemical (accessed 7 December 2017).

Chapter 10.

Metabolic engineering and synthetic biology for a bioeconomy

Metabolic engineering and synthetic biology are the core platform technologies relevant to "replacing the oil barrel". As it stands, both technologies have proven successful in basic science and in laboratory-scale applications. Their translation into bioeconomy products to date has been limited, however, often for technical reasons. This chapter identifies some of the successes, but also highlights the areas where governments could fund pre-competitive and near-market research to increase the rate of success in commercialisation. A bioeconomy presents a large conundrum, creating competition for biomass between food and industrial production. The chapter also examines the biotechnology of industrial production of bio-based materials. Ethanol, while important, is not a specific focus. A recurring theme is the need for systems integration of computational and experimental approaches, a key message for policy makers.

Introduction

There has been massive acceptance and uptake of metabolic engineering by the research community involved in bio-based applications. The research-scale successes number at least in the hundreds. Most of these syntheses are not at commercial scale; many may never make it for technical and commercial reasons; and investment in biotechnology generally lacks substantial return (Alberts et al., 2014). This section will briefly review the relevant technologies and the technical barriers to further implementation at commercial scale. These are important for future research and development (R&D) subsidy, but also for integration into innovation policy i.e. to move beyond pre-competitive research.

Technical barriers to bio-based production

Major investment in the development and deployment of efficient biomass conversion technologies is necessary (Hellsmark et al., 2016). Considerable technical barriers must be overcome to achieve a significant bio-based production industry. Some relate to the recalcitrance of cellulose and lignocellulose in the preferred feedstocks (essentially waste materials) for second-generation ethanol and bio-based materials production. Another relates to the fact that microorganisms did not evolve for operation in bioreactors at high substrate concentrations. A third revolves around the conditions of industrial operations e.g. pH extremes, solvent tolerance (see Burk and Van Dien, 2016).

Pre-treatment of biomass

The pre-treatment of biomass to degrade complex biological polymers to fermentable sugars is probably essential. It can satisfy an often-cited policy goal of using non-food crops as feedstocks for bio-production. This is specifically to avoid conflict with the policy goal of food security.

One of the more significant challenges in using the vast global lignocellulose resource is the need for large quantities of glycoside hydrolase enzymes to efficiently convert lignocellulose, hemicellulose and cellulose into fermentable sugars. The presence of lignin and hemicellulose reduces the efficiency of the biomass pre-treatment (Sun and Cheng, 2002). However, much progress has been made in the last decade in modifying enzymes. These enzymes represent the second highest contribution to raw material cost after the feedstock itself (Klein-Marcuschamer et al., 2010).

An efficient biomass degradation system likely requires a large number of enzymes to act in a co-ordinated fashion, and yet the individual and collective actions of these enzymes are poorly understood. In the consolidated bioprocessing (CBP) approach, these enzyme activities are combined with the machinery for making bio-based products within a single bacterial biocatalyst.

CBP could potentially improve bioprocess economics (Lynd et al., 2005) by avoiding the costs of a dedicated enzyme generation step. The US Department of Energy endorsed the view that CBP technology is widely considered the ultimate low-cost configuration for cellulose hydrolysis and fermentation (US DOE, 2006). Moreover, in the CBP strategy, cellulosic and hemicellulosic materials should be fermented simultaneously (see review by Hasunuma et al., 2013). In a recent example, an *E. coli* strain was engineered to express recombinant xylanases and polyhydroxyalkanoate (PHA)-producing enzymes. This was achieved for the biosynthesis of the co-polymer poly(lactate-co-3-hydroxybutyrate) [P(LA-co-3HB)] from xylan as a consolidated bioprocess (Salamanca-Cardona et al., 2016). This latter is a research success, but not yet a viable commercial process.

As another important factor, xylose is the second most-abundant carbohydrate in nature and its commercial fermentation to ethanol could provide an alternative fuel source for the future (Jeffries, 2006). The commercial yeast *Saccharomyces cerevisiae* has several advantages over most bacteria as an industrial production strain. However, *Saccharomyces cerevisiae* does not naturally use xylose as a substrate (Toivari et al., 2004) and must be engineered to both transport and ferment xylose. Several have claimed breakthroughs in the metabolic engineering of yeast to unleash this resource (e.g. Wei et al., 2013), but many challenges remain to achieve commercial viability (Moysés et al., 2016).

Nevertheless, Agbor et al. (2014) have claimed there is an increasing trend towards CBP as a path to low-cost biorefining from biomass; of the various possible bioprocessing technologies, CBP may be the most economical in the long run, but productivity is still lacking (Kawaguchi et al., 2016).

Inhibitory compounds in CBP

CBP-enabling microorganisms encounter a variety of toxic compounds produced during biomass pre-treatment that inhibit microbial growth and ethanol yield (Hasunuma and Kondo, 2012). However, the harsh conditions in the pre-treatment of the raw material release fermentation inhibitors. These include weak organic acids (particularly acetic and formic acids), furan derivatives and phenolic compounds (e.g. Almeida et al., 2007).

Several strategies have attempted to overcome the effect of inhibitors to improve fermentation ability of industrial yeast strains for ethanol production. These include controlling inhibitor concentrations during the fermentation (Martin et al., 2007); a mutagenesis and genome shuffling approach (Zheng et al., 2011); and the overexpression of genes encoding enzymes that confer resistance towards specific inhibitors (Hasunuma and Kondo, 2012). A relatively new approach to engineering tolerance to inhibitors, called global tolerance engineering, engineers a phenotype of broad tolerance towards several important inhibitors even when they are structurally dissimilar. At the same time, it keeps the number of genes being manipulated to a manageable sum (Chen and Dou, 2016).

Growth on C1 compounds

Lanzatech has pioneered the conversion of hydrogen and carbon monoxide (CO) into bio-based products at the demonstrator phase. Progress has been slow because bacteria known to use C1 substrates can be difficult to work with in an industrial setting, and many have limited genetic tools. Introduction of carbon utilisation pathways from such strains into a tractable host, such as *E. coli*, also presents significant challenges (Burk and Van Dien, 2016). Nevertheless, many C1 compounds are available in large volumes (e.g. methanol). Others are greenhouse gases that can be harnessed i.e. carbon capture and use (e.g. methane, CO_2). The low-cost and ready availability of these molecules make them attractive feedstocks for bioprocessing. Being able to harness the unique catabolic pathways, either in the native or heterologous (engineered) host could open new possibilities for non-food-based renewable feedstocks. At the same time, it could help with climate change mitigation, a primary policy goal of a bioeconomy. Technologies already exist for capturing industrial CO_2: this is used, for example, to carbonate soft drinks. But the volumes of CO_2 are tremendous compared to the volumes used in industrial processes.

Computational enzyme design

Approaches for engineering enzymes for improved activity and specificity are semi-rational at best. Although the field is still in its infancy, computational protein design could facilitate rational protein engineering or even design completely novel functions. Thousands of naturally occurring enzymes have been identified and characterised. However, numerous important applications still exist for which there are no biological catalysts that can perform the desired chemical transformation (Mak and Siegel, 2014). The ability to design specific bespoke enzymes for any given pathway step would greatly accelerate the pace of strain engineering. This, in turn, could expand opportunities to create entirely new synthetic metabolic pathways (Burk and Van Dien, 2016).

A frequently encountered challenge relates to the test component of the engineering cycle: the computer can generate many more virtual enzymes than can be tested in reality. Obexer et al. (2016) showed that microfluidic-based screening using fluorescence-activated droplet sorting is ideally suited for efficient optimisation of designed enzymes with low starting activity, essentially straight out of the computer.

This is a recurring message for policy makers in R&D subsidy – the ability to design and build often exceeds the ability to test. The answers lie in biology, but with system integration and automation. The frontier is in integrated computational/experimental metabolic engineering platforms to design, create and optimise novel high-performance enzymes, but also organisms and bioprocesses (Barton et al., 2015). As the data sets become larger, the systems biology approach will become essential rather than the exception.

Minimal cells for bio-contained microbial factories

The start point for designing future production strains will be minimal, or chassis cells. These are self-replicating minimal machines that can be tailored to produce specific chemicals or fuels. These machines will remove non-essential energy-consuming pathways and carbon sinks, and minimise regulatory and toxicity issues (Vickers et al., 2010; Lee et al., 2013). To date, constraints have shaped the regulatory/metabolic pathways of the production strain. In future, it will be less effort to construct functional "circuits" from scratch (Ghim et al., 2010). This is coming closer as the price of DNA synthesis has tumbled.

Ostrov et al. (2016) have made a significant advance towards a chassis *E. coli* industrial production strain. The decreasing cost of DNA synthesis has greatly reduced the financial barriers to synthesising entire genomes. They have developed computational and experimental tools to rapidly design and prototype synthetic organisms. As much as synthetic genomes have already been reported, this effort is on a scale that has not yet been explored.

The ultimate aim of this work is to produce a virus-resistant, bio-contained bacterium for industrial applications. Once complete, their genetically isolated *rE.coli-57* will offer a unique chassis with expanded synthetic functionality that will be broadly applicable for biotechnology. At current costs, the project could be attained for around USD 1 million.

Biocontainment to prevent escape of genetically modified microbes into the environment remains another goal for industrial production strains. There are necessary, but insufficient, metrics to evaluate biocontainment (Mandell et al., 2015), and therefore design strategies are incomplete. Progress is continuous, but no single existing mechanism can guarantee biocontainment.

Small-scale fermentation models

Fermenters are the ultimate arbiters of process optimisation, but are costly to run and typically require expert supervision. Multiplexing the design and test process should drastically reduce the number of strains to be tested. However, the fermenter is still necessary to ensure the truly best strains are chosen for industrial production. Small-scale fermentation is lacking in a number of areas, such as pH and aeration control and the ability to sample frequently. Solutions will hopefully lie in microfluidics (Churski et al., 2015; Burk and Van Dien, 2016).

Robustness

Tolerance to inhibitors is part of robustness of microbial production strains. Natural microorganisms were not intended for the conditions of industrial production. As a result, they must be engineered with new characteristics to make them more robust, a classic function for synthetic biology (Zhu et al., 2012). In an industrial bioreactor, where nutrient levels are often in excess, the environment constantly changes (Wang and Zhong, 2007). This process may produce toxic metabolites. High levels of shear stress may be applicable in a bioreactor (Chisti, 1999).

There are still only a few examples that have deliberately employed synthetic biology to increase robustness in bio-based production. For instance, butanol offers some advantages over ethanol as a biofuel (e.g. Abdehagh et al., 2014). However, low yield and titre in the fermentative clostridia hinder development of butanol (Xue et al., 2013). DARPA has introduced a research programme dedicated to robustness in synthetic biology (Box 10.1).

Box 10.1. **DARPA (United States) and the Biological Robustness in Complex Settings (BRICS) programme**

The Biological Robustness in Complex Settings (BRICS) programme seeks to develop the fundamental understanding and component technologies needed to engineer bio-systems that function reliably in changing environments. A long-term goal is to enable the safe transition of synthetic biological systems from well-defined laboratory environments into more complex settings. In this environment, they could achieve greater biomedical, industrial and strategic potential.

To date, work in synthetic biology has focused primarily on manipulating individual species of domesticated organisms. These species tend to be fragile – they require precise environmental controls to survive, and can lose their engineered advantages through genetic attrition or recombination. The costs of maintaining required environmental controls, and detecting and compensating for genetic alterations are substantial.

The BRICS portfolio will consist of programmes to elucidate the design principles of engineering robust biological consortia. It will apply this fundamental understanding towards specific applications e.g. on-demand bio-production of novel drugs, fuels and coatings.

Source: Adapted from DARPA (n.d.), "Biological Robustness in Complex Settings", www.darpa.mil/program/biological-robustness-in-complex-settings.

Titre, yield and productivity

Most natural microbial processes are incompatible with an industrial process; the product titres (g per litre of product), yields (g product per g substrate, often glucose) and productivity (g per litre per hour) rates are often too low to be scalable (e.g. Lee et al., 2013; Harder et al., 2016). The required economic yield, titre and productivity of a microbial process depend on whether the product is a bulk or niche chemical. Higher values of the chemical can increasingly tolerate low titre, yield and productivity. For low-value, bulk chemicals, however, these factors make or break a bioprocess.

A fundamental constraint on host cell productivity is the metabolic burdens that lead to undesirable physiological changes. Engineering cell metabolism for bio-production not only consumes building blocks and energy molecules such as ATP, but also triggers energetic inefficiencies in the cell (Wu et al., 2016). The authors stated that requirements for higher success rates in industrial settings calls for novel genome-scale models, ^{13}C-metabolic flux analysis and machine learning for weighting, standardisation and predicting metabolic costs.

Gene and genome editing in production strains

Despite recent advances, the sheer size of even the smallest bacterial genomes renders serial modification of limited utility for truly genome-scale engineering endeavours. Targeted genome editing and engineering have until recently been laborious and costly. Efficient methods enabling multiplex genome editing are urgently needed (Esvelt and Wang, 2013). Here, progress has been rapid even in the last three years. In combination with more sophisticated metabolic modelling tools, such new techniques will substantially accelerate metabolic engineering (Sandoval et al., 2012). This could transform the costs involved in making new strains for bio-based production. Chromosomal insertion and gene editing have particular potential in Saccharomyces cerevisiae, the "original" industrial microbe used for ethanol production.

Conclusions

The story remains the same for bio-based production through metabolic engineering – from Herculean research efforts and cash burn, to many ideas from new molecules to target chassis strains and biocontainment strategies. There are large numbers of academic groups with specific interests in individual areas of research. However, relatively few academic groups or companies can bring together a commercial project from idea to finished product ready for a bioprocess.

R&D subsidy decision makers need to rethink R&D programmes in the area. Challenges across the board are demanding ever-larger data sets to shrink the number of actual physical experiments to reasonably attainable levels. The message to these decision makers is clear: it is now time to fund computational and experimental systems integration. Otherwise, standard methodologies and interoperability will become more distant, not less.

References

Abdehagh, N. et al. (2014), "Separation techniques in butanol production: Challenges and developments", *Biomass and Bioenergy*, Vol. 60, Elsevier, Amsterdam, pp. 222-246.

Agbor, V. et al. (2014), "Biomass pretreatment for consolidated bioprocessing (CBP)" in *Advances in Biorefineries: Biomass and Waste Supply Chain Exploitation*, K. Waldron (ed.), Woodhead Publishing Limited, UK.

Alberts, B. et al. (2014), "Rescuing US biomedical research from its systemic flaws", *Proceedings of the National Academy of Sciences*, Vol. 111/16, pp. 5773-5777.

Almeida, J.R.M. et al. (2007), "Increased tolerance and conversion of inhibitors in lignocellulosic hydrolysates by *Saccharomyces cerevisiae*", *Journal of Chemical Technology and Biotechnology*, Vol. 82/4, Wiley Online Library, pp. 340-349.

Barton, N.R., et al. (2015), "An integrated biotechnology platform for developing sustainable chemical processes", *Journal of Industrial Microbiology & Biotechnology*, Vol. 42/3, Springer Link, pp. 349-360.

Burk, M.J. and S. Van Dien (2016), "Biotechnology and chemical production: Challenges and opportunities", *Trends in Biotechnology*, Vol. 34, Elsevier, Amsterdam, pp. 187-190.

Chen, R. and J. Dou (2016), "Biofuels and bio-based chemicals from lignocellulose: Metabolic engineering strategies in strain development", *Biotechnology Letters*, Vol. 38/2, Springer Link, pp. 213-221.

Chisti, Y. (1999), "Shear sensitivity", in *Encyclopedia of Bioprocess Technology: Fermentation, Biocatalysis and Bioseparation, Vol. 5*, Flickinger, M.C. and S.W. Drew (eds.), Wiley, New York, pp. 2379-2406.

Churski, K. et al. (2015), "Droplet microfluidic technique for the study", *Micromachines*, Vol. 6/10, MPDI, Basel, pp. 1514-1525.

DARPA (n.d.), "Biological Robustness in Complex Settings", webpage, www.darpa.mil/program/biological-robustness-in-complex-settings (accessed 8 January 2018).

Esvelt, K.M. and H.H. Wang (2013), "Genome-scale engineering for systems and synthetic biology", *Molecular Systems Biology,* Vol. 9/1, Embo Press, Heidelberg, pp. 641, http://dx.doi.org/10.1038/msb.2012.66.

Ghim, C.-M., et al. (2010), "Synthetic biology for biofuels: Building designer microbes from the scratch", *Biotechnology and Bioprocess Engineering*, Vol. 15, Springer Link, pp. 11-21.

Harder, B-J. et al. (2016), "Model-based metabolic engineering enables high yield itaconic acid production by *Escherichia coli*", *Metabolic Engineering*, Vol. 38, Elsevier, Amsterdam, pp. 29-37.

Hasunuma, T. and A. Kondo (2012), "Development of yeast cell factories for consolidated bioprocessing of lignocellulose to bioethanol through cell surface engineering", *Biotechnology Advances*, Vol. 30, Elsevier, Amsterdam, pp. 1207-1218.

Hasunuma, T. et al. (2013), "A review of enzymes and microbes for lignocellulosic biorefinery and the possibility of their application to consolidated bioprocessing technology", *Bioresource Technology*, Vol. 135, Elsevier, Amsterdam, pp. 513-522.

Hellsmark, H. et al. (2016), "Innovation system strengths and weaknesses in progressing sustainable technology: The case of Swedish biorefinery development", *Journal of Cleaner Production*, Vol. 131, Elsevier, Amsterdam, pp. 702-715.

Jeffries, T.W. (2006), "Engineering yeasts for xylose metabolism", *Current Opinion in Biotechnology*, Vol. 17, Elsevier, Amsterdam, pp. 320-326.

Kawaguchi, H. et al. (2016), "Bioprocessing of bio-based chemicals produced from lignocellulosic feedstocks", *Current Opinion in Biotechnology*, Vol. 42, Elsevier, Amsterdam, pp. 30-39.

Klein-Marcuschamer, D. et al. (2010), "Techno-economic analysis of biofuels: A wiki-based platform for lignocellulosic biorefineries", *Biomass and Bioenergy*, Vol. 34/12, Elsevier, Amsterdam, pp. 1914-1921.

Lee, S.L. et al. (2013), "Design and development of synthetic microbial platform cells for bioenergy", *Frontiers in Microbiology*, Vol. 4, Frontiers Media, Lausanne, pp. 92, http://dx.doi.org/10.3389/fmicb.2013.00092.

Lynd, L.R. et al. (2005), "Consolidated bioprocessing of cellulosic biomass: An update", *Current Opinion in Biotechnology*, Vol. 16/5, Elsevier, Amsterdam, pp. 577-583.

Mak, W.S. and J.B. Siegel (2014), "Computational enzyme design: Transitioning from catalytic proteins to enzymes", *Current Opinion in Structural Biology*, Vol. 27/1, Elsevier, Amsterdam, pp. 87-94.

Mandell, D.L. et al. (2015), "Biocontainment of genetically modified organisms by synthetic protein design", *Nature*, Vol. 518/7537, Nature Publishing Group, London, pp. 55-60.

Martin, C. et al. (2007), "Dilute sulfuric acid pretreatment of agricultural and agro-industrial residues for ethanol production", *Applied Biochemistry and Biotechnology*, Vol. 137, Springer Link, pp. 339-352.

Moysés, D.N. et al. (2016), "Xylose fermentation by *Saccharomyces cerevisiae*: Challenges and prospects", *International Journal of Molecular Sciences*, Vol. 17, MPDI, Basel, pp. 207, http://dx.doi.org/10.3390/ijms17030207.

Obexer, R. et al. (2016), "Efficient laboratory evolution of computationally designed enzymes with low starting activities using fluorescence-activated droplet sorting", *Protein Engineering, Design and Selection*, Vol. 30/7, Oxford University Press, pp. 355-366.

Ostrov, N. et al. (2016), "Design, synthesis and testing toward a 57-codon genome", *Science*, Vol. 353/6301, American Association for the Advancement of Science, Washington, DC, pp. 819-822.

Salamanca-Cardona, L. et al. (2016), "Consolidated bioprocessing of poly(lactate-co-3-hydroxybutyrate) from xylan as a sole feedstock by genetically-engineered *Escherichia coli*", *Journal of Bioscience and Bioengineering*, Vol. 122/4, Elsevier, Amsterdam, pp. 406-414.

Sandoval, N.R. et al. (2012), "Strategy for directing combinatorial genome engineering in *Escherichia coli*", *Proceedings of the National Academy of Sciences*, Vol. 109, pp. 10540-10545.

Sun, Y. and J. Cheng (2002), "Hydrolysis of lignocellulosic materials for ethanol production: A review", *Bioresource Technology*, Vol. 83/1, Elsevier, Amsterdam, pp. 1-11.

Toivari, M.H. et al. (2004), "Endogenous xylose pathway in *Saccharomyces cerevisiae*", *Applied and Environmental Microbiology*, Vol. 70/6, American Society for Microbiology, Washington, DC, pp. 3681-3686.

US DOE (2006), "Breaking the biological barriers to cellulosic ethanol: A joint research agenda", DOE/SC-0095, US Department of Energy Office of Science and Office of Energy Efficiency and Renewable Energy, Washington, DC, www.doegenomestolife.org/biofuels.

Vickers, C.E. et al. (2010), "Chassis cells for biochemical production", *Nature Chemical Biology*, Vol. 6/12, Nature Publishing Group, London, pp. 875-877.

Wang, S.J. and J.-J. Zhong (2007), "Bioreactor engineering", in *Bioprocessing for Value-Added Produces from Renewable Resources, New Technologies and Applications*, S-T Yang (ed.), Elsevier, Amsterdam.

Wei, N. et al. (2013), "Enhanced biofuel production through coupled acetic acid and xylose consumption by engineered yeast", *Nature Communications*, Vol. 4/2580, Nature Publishing Group, London, http://dx.doi.org/10.1038/ncomms3580.

Wu, G. et al. (2016), "Metabolic burden: Cornerstones in synthetic biology and metabolic engineering applications", *Trends in Biotechnology*, Vol. 34/8, Elsevier, Amsterdam, pp. 652-664.

Xue, C. et al. (2013), "Prospective and development of butanol as an advanced biofuel", *Biotechnology Advances*, Vol. 31, Elsevier, Amsterdam, pp. 1575-1584, http://dx.doi.org/10.1016/j.biotechadv.2013.08.004.

Zheng, D.Q. et al. (2011), "Screening and construction of *Saccharomyces cerevisiae* strains with improved multi-tolerance and bioethanol fermentation performance", *Bioresource Technology*, Vol. 102, Elsevier, Amsterdam, pp. 3020-3027.

Zhu, L. et al. (2012), "Engineering the robustness of industrial microbes through synthetic biology", *Trends in Microbiology*, Vol. 20/2, Elsevier, Amsterdam, pp. 94-101.

Chapter 11.

Education and training for industrial biotechnology

This chapter examines education and training for industrial biotechnology, a field that calls for education outside of normal disciplinary boundaries. Many factors in the education and training of industrial biotechnologists point to multi-disciplinarity. This has been discussed many times, but has been elusive in practice. The most obvious combination of skills needed is synthetic biology or genetic engineering with "green" chemistry, with the reduction-to-practice skills provided by chemical engineering. Other mathematical skills are also important. But for employment in small companies, employees also need to be flexible and willing to multi-task and get soft tasks done. This often does not suit a PhD graduate as doctoral training remains specialist, long-term and driven by publication. Although these issues could have been part of a capacity building discussion in this book, the significant policy implications warrant their own chapter.

Introduction

OECD analysis suggests that innovation thrives in an environment characterised by a few key features, including a skilled workforce (OECD, 2015). Skills will be central to enabling bio-production due to the newness of the subject, its multi-disciplinary nature, the complexity of biology and bio-production, and the need for many stakeholders with different skills. Jobs for the workforce, not only research jobs, are a major goal of bio-production. This will only work well by rethinking education.

A key discipline of industrial biotechnology is microbiology. Life sciences PhD-level education remains focused on training for academic careers (American Society for Microbiology, 2013). However, data from 2010 published in the National Science Board (NSB) 2014 Science and Engineering Indicators show that a mere 29% of newly graduated life science PhDs will find a full-time faculty position in the United States (Figure 11.1). A recent review also confirmed that growth in the number of US post-doctoral researchers far exceeds the growth in the number of tenure-track job openings (National Academy of Sciences, 2014a). There are simply too many PhD students and too few senior posts (Nature, 2016). On the positive side, then, post-graduates should have plenty of choices for entry into industrial biotechnology. However, in microbiology, the field is overwhelmingly dominated by medical microbiology.

Figure 11.1. **Likelihood to work in academia for newly graduated PhDs**

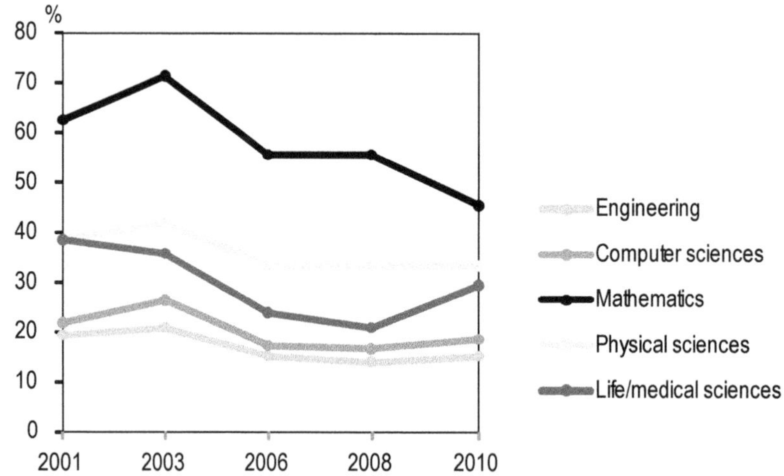

Note: The low figures for engineering and computer sciences reflect the greater likelihood that these PhD graduates will enter industry.

Source: Delebecque and Philp, unpublished data.

The problems are far from new. As far back as 1995, a report expressed the need for change in the education of scientists and engineers (National Academy of Sciences, 1995). This report was concerned that the United States was producing too many PhDs. It said that industry often complained that graduates were too specialised to accomplish the range of tasks they would be confronted with. Also, when scientists form small biotechnology companies, they are often placed in a managerial role in which they may have no training or know-how (Corolleur et al., 2004). This has all brought about a call for a new type of PhD, one that offers much more breadth and flexibility.

The challenge of multi-disciplinary education

Traditional scientific education and training has remained divided by disciplines such as microbiology, chemistry and computing. The long-standing conundrum of multi-disciplinary education is the need for both breadth and depth. The challenge to higher education remains on many levels. For example, a central theme in bioeconomy strategies is sustainability. Training in sustainability itself begs multi-disciplinarity as some of the depth skills needed are systems thinking, strategic planning, and evaluating environmental, social and economic performance. This educational conundrum for sustainability (Mascarelli, 2013) is the same for industrial biotechnology: how to make the inter-disciplinary approach not only substantive, but also practical for early-career scientists.

Life sciences industry-wide issues

Some of the life sciences-wide industry issues are clearly crystallised in an American report (CBSI, 2013). Employer interviews identified several industry-wide gaps in the capabilities and talent of the workforce pool and proposed reasons:

- The life science industry has a decreased need for deeply trained senior scientists. There is an over-specialisation surplus, whereas employers are looking for a workforce with greater breadth and more soft skills.
- Academic programmes are training students by discipline and not by problem-solving, which typically requires cross-disciplinary skills and capabilities.
- Apprenticeships and long-term training programmes are lacking.

The UK biopharmaceuticals industry recently highlighted major skills gaps in mathematical and computational areas. These have emerged due to the rapid development of new disciplines such as systems biology and health informatics (ABPI, 2015). For the industrial biotechnology industry, the same holds true. The following sections address some specific and critical training gaps to foster industrial biotechnology and synthetic biology-based manufacturing. They attempt to examine the future look of the workforce and related research base if this activity gathers momentum in response to societal grand challenges. This may guide governments in directions for higher education.

The critical workforce gaps in bio-based manufacturing

Finding biologists is not the most difficult task for bio-based manufacturing. Automation engineers specialising in high-throughput strain production critical to synthetic biology-based manufacturing are rare. Managing automated systems will have to be a skill set for graduates in biology and chemistry in the future (Extance, 2016). For a long time, it has also been difficult to find fermentation staff: this is the province of the biochemical engineer, who combines the mathematics of cell growth with bioreactor and bioprocess design. And yet bio-manufacturing is the common operation that links together all the different market sectors of the world's biotechnology industry.

Perhaps hardest to find of all are employees well versed in experimental design and statistics, especially now that large data sets are becoming more common. Big data is creating an imperative for more complex design that enables fewer experiments and trials. Scientific irreproducibility – the inability to repeat others' experiments and reach the same conclusion – is a growing concern. Yet few early-career researchers receive formal instruction on topics like experimental design and flaws in statistical analysis (Baker, 2016).

This diverse group of employees is essential for a functional synthetic biology-based production plant, but remains rare as this business sector is a small niche. As sector growth is difficult to forecast, it challenges governments to predict how to invest in and reform higher education to create a workforce that matches the growth dynamics of the sector.

Bioinformatics may be a major roadblock

The bottleneck for the growing industrial biotechnology industry is shifting to bioinformatics and data mining. Data mining tools akin to the ones revolutionising social sciences and linguistics will become essential. The Short Read Archive at the US National Center for Biological Information is set to exceed a petabyte (National Academy of Sciences, 2013). As high-throughput sequencing is increasingly deployed across research organisations, hospitals, biotechnology facilities and companies, the acquisition of genomic information will also burgeon. DNA synthesis costs have tumbled between 2014 and 2016. These price decreases, combined with advances in next-generation sequencing, are increasing the need and role of advanced software design tools.

"Dry lab" skills have traditionally been isolated from "wet lab" ones. Nevertheless, bioinformatics requires deep knowledge of biology theory and mathematics/computing. These fields are usually not taught in depth in the same programme in higher education, and this is but one more challenge to be overcome.

The scientist as engineer

Engineering education depends on several key concepts that have been largely missing in biotechnology (Panke, 2008): comprehensiveness of available relevant knowledge; orthogonality; hierarchy of abstraction; standardisation; and separation of design from manufacture. Systems modelling and design are well-established in engineering disciplines, but until recently have been rare in biology. The sheer complexity of biology has also hindered development of its formal mathematics. Synthetic biology has started to bridge the gap between biology and engineering (Liu, Hoynes-O'Connor and Zhang, 2013).

The education of a biologist, which still focuses more on the needs of research, has been dominated by a more descriptive tradition. This contrasts with engineering, dominated by a much more quantitative tradition, and the need to standardise and reduce complexity to practice.

However, this comes at a time of widely conflicting attitudes to engineering education. For example, only 4.4% of the undergraduate degrees awarded by US colleges and universities are in engineering. This compares with 13% of similar degrees awarded in key European countries and 23% in key Asian countries (National Academy of Sciences, 2014b).

With the continuing relationship between technology and discovery, Botstein (2010) contends that cell biologists in the next 50 years will have to be conversant with a broader range of concepts. This will range from physics through chemistry to genetics. However, they will especially need to know mathematical and computational methodologies that drive technology development.

The quantitative theoretical and computational component represents a fundamental departure from the tradition of the life sciences. Nevertheless, Tadmor and Tidor (2005) stressed that modelling should not be construed as a replacement for experimentation. Indeed, large stores of practical and theoretical knowledge are essential for one to function in a laboratory environment. But creating this depth of laboratory skills is among the most expensive and time-consuming elements of higher education. It leads back to the dilemma of breadth versus depth versus adaptability.

The chemical engineer as a role model?

Chemical engineers have played a tremendous role in generating and transferring the enormous benefits of the chemicals industry to society. The mathematics and thermodynamics of chemical engineering enabled the transfer of chemistry from the laboratory to full-scale industrial production, using crude oil as the raw material. For industrial biotechnology to fulfil its promise in a bioeconomy, these skills will be essential, with the new raw material being biomass. Chemical and biochemical engineers are key elements of the future bioeconomy because they alone can set the production agenda, knowing the process, energy, materials and cost elements (Woodley et al., 2013).

The chemical engineering curriculum is already full. Chemical engineering students may not wish to have industrial biotechnology in the undergraduate curriculum. Such a move could divert them from their main objective: to get certification to practice in the chemical and petrochemical industries. This may indicate a niche for training chemical engineers at Master's level in industrial biotechnology.

Synthetic biology education: Another key "inter-discipline"

The education system has been responding to the needs of the growing synthetic biology community. The number of courses in synthetic biology has grown at a tremendous rate, with at least 100 institutions involved (Delebecque and Philp, 2015). However, many do not focus on industrial production. Therefore, industrial biotechnology courses, and organisations teaching them, are still very much pioneers.

Beyond science and engineering

Given the history of the genetic modification (GM) debate, such matters as public perception will also shadow industrial biotechnology and synthetic biology. There is already evidence that political and economic pressures will guide development of synthetic biology (e.g. Rai and Boyle, 2007). Kuldell (2007) argued that educational efforts that fail to equip students for these aspects of the emerging discipline are unsound. Public engagement, though fraught, is necessary for the acceptance of synthetic biology and industrial biotechnology more broadly. Public engagement is weakened by a lack of a standard approach (National Academy of Sciences, 2016). Policy makers could include social scientists and ethicists in strategies for developing and encouraging the uptake of bio-based products, and have this embedded in education. On the other hand, public engagement should not become a "mode of governance" of research (Kuntz, 2016).

To make employees fit for the workplace, this education also needs to encompass other practices such as regulatory compliance, risk assessment and biosafety, and good manufacturing practice (GMP). These practices are not academic research disciplines. But they can change rapidly, with far-reaching consequences for a small company. In-house training in GMP, for example, takes up considerable time and human resources. It can be a burden for small companies.

The many faces of regulation

Bio-based production creates regulatory challenges across boundaries as well. The metabolically engineered microbes (i.e. process) are subject to GMO regulation. At the same time, the chemicals and fuels (i.e. products), often being drop-in substitutes for fossil-derived materials, are subject to chemical regulations. These include the Toxic Substances Control Act (TSCA) in the United States and Registration, Evaluation, Authorisation and Restriction of Chemicals (REACH) in Europe.

Some approaches to industrial biotechnology education and training

The US NSF Center for Biorenewable Chemicals, a third generation Engineering Research Center (ERC) established at Iowa State University in 2008, is a classic approach to industrial biotechnology education and training. It includes five core partner universities, two affiliated research centres, four international partner institutions, and multiple industrial partners and pre-college entities (CBIRC, n.d.). The ERC's mission is based on research and education principles that seek to transform the existing petroleum-based chemical industry into an industry based on renewables (Haen et al., 2012). It offers courses for school teachers, through undergraduate and graduate education.

Undergraduate courses: Preparing the way

It is probably too early for entire undergraduate degrees to train biologists in industrial manufacturing. However, relevant science undergraduate degrees could be re-designed to serve as a platform for post-graduate study. For example, one of the key disciplines, microbiology, has curricula overwhelmingly dominated by medical microbiology. A re-orientation of microbiology undergraduate education could include, apart from the core microbiological skills, quantitative skills that are important for success in industry. Students so equipped with skills in calculus, linear algebra, statistics, large dataset management and programming (American Society for Microbiology, 2013) would be better disposed to embed industrial aspects of biotechnology.

In Canada, two universities are strengthening undergraduate programmes in biotechnology. The faculties of science (biochemistry) and engineering (chemical engineering) at the University of Ottawa jointly offer an undergraduate biotechnology programme. The University of Guelph offers an undergraduate programme in biological engineering that focuses on fundamentals in biomaterials science, bio-systems analysis, bio-mechanics, instrumentation and digital control. The programme can be tailored to explore interests in the production of renewable fuels such as ethanol and biodiesel; sustainable bioplastics made from plant materials; the extraction and stabilisation of nutraceuticals to provide health benefits; or the manufacturing of safe food products.

Taught and Research Master's

Industrial biotechnology lends itself well to a research Master's degree, emphasising practice-led research combined with relatively few taught modules compared with other graduate degrees. This sort of degree is designed in most cases to prepare students for doctoral research. However, it is also useful for those considering a career in the private sector where research is a key focus, but a PhD is not specifically required.

Various institutions around the world have begun to offer graduate degrees related to industrial biotechnology. The University of Georgia Master of Biomanufacturing and Bioprocessing degree (UGA, n.d.), a two-year programme, claims a unique focus on the full bio-manufacturing experience with hands-on training and exposure to industrial grade equipment. Its curriculum includes academic courses in science (e.g. biofuels/biochemical, pharmaceuticals manufacture) and business (e.g. finance, supply chain issues and manufacturing practices). It also offers professional training with cutting edge companies through case study projects and internships. Instead of producing a traditional thesis, students complete a research project during the summer of year one and a 400-hour industry internship during the summer of year two.

The La Trobe University (Australia) Master of Biotechnology and Bioinformatics focuses on the interface of molecular biology and information technology. It uses the power of computing to tackle biological and medical problems. The need for bioinformatics graduates will increase as computational tools are increasingly incorporated into bio-production.

The University of Cagliari (Italy) Master in Chemical and Biotechnological Process Engineering combines the skills of chemical engineering with the needs of the biotechnology industry. A goal is to teach students how to use the increased knowledge of chemical, physical and biological sciences to develop advanced mathematical models for chemical and biotechnological processes.

The Grenoble Ecole de Management (France) Master Specialised Management of Biotech Companies aims to provide specific managerial skills and understanding of issues related to the sector, as well as training in change management and the specific challenges of the biotechnology sector.

The University of Guelph, Ontario Master of Biotechnology programme brings together the Department of Molecular & Cellular Biology, and the Department of Management to offer courses in business skills (e.g. commercialising innovations) in addition to deeper scientific training.

Massive open online courses

The traditional on-campus experience could be radically changed by the explosion of massive open online courses (MOOCs), which will enhance classroom and laboratory work. The evidence for the impact of MOOCs is still embryonic. More analysis is needed as greater experience is acquired with their use. A number of MOOC platforms, such as Coursera (Coursera, n.d.) and edX (edX, n.d.), now propose a wide array of classes spanning engineering to molecular biology and all the building blocks in between that can provide the basic toolset to start practising engineering biology. A specialist MOOC for industrial biotechnology is offered jointly by the Technical University of Delft and the University of Campinas (Box 11.1).

Box 11.1. **edX course in industrial biotechnology**

The edX course in industrial biotechnology is a joint initiative of TU Delft (Netherlands), the international BE-Basic consortium and University of Campinas (Brazil). It provides the insights and tools for the design of sustainable biotechnology processes. Students use the basics of industrial biotechnology for design of fermentation processes to produce fuels, chemicals and foodstuffs (BE-Basic Foundation, 2016). Throughout this course, students are challenged to design a biotechnological process and evaluate its performance and sustainability.

Combining edX with other relevant courses can build the broader education that bioeconomy and industrial bio-based manufacturing seems to need. For example, TU Delft offers another MOOC course on responsible innovation. This discipline considers new technologies that are being developed in response to social challenges (e.g. food safety, smart cities, sustainable energy and digital security).

The TU Delft MOOCs are offered through the online edX platform, where MIT, Harvard and other universities have been making courses available to anyone with an internet connection since 2012. TU Delft chose to use edX partly because the platform allows publication of materials with an open licence, making it possible for others to use the materials.

Arguably, MIT pioneered online learning, building on research that consistently showed that students perform better when they take both traditional and online courses than when they take only one (National Academy of Sciences, 2014b). The MIT and Harvard-owned edX MOOC platform differs from other such platforms as it is non-profit and runs on open-source software. Unlike the traditional lecture, each lesson is a ten-minute video on a single concept followed by self-assessment tools.

MOOCs are easily scalable and adaptable, which are important benefits. Industrial biotechnology is expanding and changing rapidly. As a result, educational materials lose their freshness, if not their relevance. When the hard foundational work of creating a MOOC is done, software and screencasts could replace or upgrade course content in a matter of minutes.

Specialist training facilities

For early-career scientists, gaining access to bio-based production experience is difficult because universities do not normally have such facilities. Ireland is an exception with its National Institutes model, which includes a dedicated facility for training in bioprocessing (the National Institute for Bioprocessing Research and Training, NIBRT). For a relatively small country, Ireland has a large pharmaceuticals sector. NIBRT provides a "one stop shop" for bioprocessing training requirements (NIBRT, n.d.). The institute builds tailored solutions for clients, ranging from operator through to senior management training. Further, it delivers training in a realistic environment with simulated good manufacturing practices (GMPs). This type of environment, not typically found in universities, is more appropriate to train industry professionals. Equally, undergraduate and graduate programmes could use such a facility to expose students to industry working conditions.

A role for intermediate research organisations and laboratories

Intermediate research organisations (IROs) can enable work in selected fields without the conflicting pressures of publishing and teaching explicit in academic research (Gauvreau, Winickoff and Philp, 2018). The concept seems enshrined in the UK Catapult model (Catapult, n.d.), a concept that could have been tailor-made for industrial biotechnology. Further, such a model has great potential to fill skill gaps (more on the apprentice, hands-on model rather than on the academic student model).

Such a model can work for PhD students as well. For example, the RIKEN Junior Research Associate programme in Japan provides part-time positions for young researchers enrolled in Japanese university PhD programmes (RIKEN, n.d.). This enables PhD students to carry out research alongside RIKEN scientists and also strengthens relationships between RIKEN and universities in Japan.

Business management education and training for the industrial biotechnology industry

One solution to a shortage of experienced managers in the biotechnology industry has been to create a specific stream for biotechnology within the normally generic MBA programme (OECD, 2005). Theories of business administration have their roots in commerce, which has in the past been focused on non-technological issues (Lambert, 2004). Therefore, the typical MBA programme is not particularly well-suited to industrial biotechnology business management. Given the pressures on small companies active in industrial biotechnology, much shorter courses that focus on specific skills gaps may be more appropriate.

Given the potential impact of industrial biotechnology on the chemicals industry, a European five-day mini-MBA[1] tailored for mid-level chemical industry managers is pertinent. These managers' roles are being impacted by the rapidly evolving trends of globalisation; registration, evaluation and authorisation of chemicals; green chemistry; waste reduction; sustainability; and operational efficiency. All are also directly relevant to the emerging bio-based industry.

A four-day synthetic biology-specific MBA has also been developed (SynbiCITE, n.d.). It covers the main strategies required to establish, build and manage a biotechnology company built around synthetic biology. It has focused on the early stages of setting up a company, getting funding and understanding the wider reaches of intellectual property.

Training technicians: Providing the bio-manufacturing workforce

Technicians are workforce employees, not researchers. As they are responsible for day-to-day work in bio-manufacturing, technicians will be required in higher numbers than researchers and need a broader range of skills. Industrial biotechnology should be part of their training, not all of it. Some foreseeable functions will be routine maintenance of metabolically engineered strains; embedding synthetic biology with GMP guidance; regulatory and compliance training e.g. bio-banking, transportation of live biologicals and document management; standard operating procedures to deal with accidental spills and releases.

They should also be trained in matters such as scale-up, knowledge of packaging and labelling protocols (Wallman et al., 2013). Scale-up is a massive technical barrier in the bio-based industry, especially at the scale of transportation fuels (Westfall and Gardner, 2011), stretching the skills of both strain and fermentation engineers.

Manufacturing does not fit well into normal boundaries of university degree programmes. As a result, it is often marginalised (Glaser, 2013). An approach that creates a vocational workforce locally and separately from the universities – in technical and community colleges, for example – would take pressure off the universities. It would also create more jobs and investment in local or community colleges. This aligns well with thinking that envisages creation of biorefineries and bioeconomy clusters in rural environments as a means of rural regeneration.

The recently established Industrial Biotechnology Innovation Centre in Scotland has developed a range of educational programmes with its collaborative partners to meet the need of the biotechnology industry (IBioIC, n.d.). Jointly with the Forth Valley College, the Higher National Diploma (HND) is specifically designed to create a cadre of technical staff. The newly developed HND involves the study of three crucial disciplines: biology, chemistry and engineering, on either a full-time or part-time basis.

Given national and international plans for biotechnology education, it would desirable to harmonise and standardise qualifications in bio-manufacturing, and probably even essential. Again, this is not necessarily what universities aim to achieve. It is best done close to manufacturing to allow efficient cross-fertilisation between training institutions and industry.

Conclusions

Industrial biotechnology and synthetic biology training requires a paradigm shift in how education is structured. Programmes are needed that encourage creativity and exploration, while harnessing the truly unique inter-disciplinary nature of the field and harvesting the different forms of training highlighted above. To keep pace in a changing world, beyond the traditional debate of depth versus breadth in education, one of the answers lies in training for adaptability and dynamism. Pioneering universities answering this challenge are, and will be, training the next generation of creative and co-operative knowledge and venture builders. These graduates will be able to update and productively use their knowledge to drive innovation. The gradual shift to biomass from crude oil and natural gas as the raw material for production will present a plethora of technical difficulties. It will demand the ability to use knowledge co-operatively to create the factories and products of the future. In response, the shift calls for equally innovative education and bold reforms.

Note

1. These courses do not allow graduates to use the initials "MBA" after their names.

References

ABPI (2015), *Bridging the Skills Gap in the Biopharmaceutical Industry: Maintaining the UK's Leading Position in Life Sciences*, Association of the British Pharmaceutical Industry, November.

American Society for Microbiology (2013), "Microbe-powered jobs: How microbiologists can help build the bioeconomy", Report on a Colloquium, American Academy of Microbiology, Dallas.

Baker, M. (2016), "Seek out stronger science", *Nature*, Vol. 537/7622, Nature Publishing Group, London, pp. 703-704.

BE-Basic Foundation (2016), "MOOC Industrial Biotechnology", webpage, www.be-basic.org/news-center/news/mooc-industrial-biotechnology.html (accessed 7 December 2017).

Botstein, D. (2010), "Technological innovation leads to fundamental understanding in cell biology", *Molecular Biology of the Cell*, Vol. 21/22, American Society for Cell Biology, Bethesda, United States, pp. 3791-3792.

Catapult (n.d.), Catapult website, https://catapult.org.uk (accessed 7 December 2017).

CBIRC (n.d.), Center for Biorenewable Chemicals website, www.cbirc.iastate.edu (accessed 7 December 2017).

CBSI (Coalition of State Bioscience Institutes) (2013), "Demand for talent: Current & projected workforce trends in the life science industry", A joint report of CBSI and Booz & Co, www.csbinstitutes.org/download/files/reports/CSBI_WorkforceReportvFR.pdf.

Corolleur, C.D.F. et al. (2004), "Turning scientific and technological human capital into economic capital: The experience of biotech start-ups in France", *Research Policy*, Vol. 33/4, Elsevier, Amsterdam, pp. 631-642.

Coursera (n.d.), Coursera website, www.coursera.org (accessed 7 December 2017).

Delebecque, C. and J. Philp (2 June 2015), "Training for synthetic biology jobs in the new bioeconomy", Science Careers blog, http://dx.doi.org/10.1126/science.caredit.a1500143.

edX (n.d.), edX website, www.edx.org (accessed 7 December 2017).

Extance, A. (17 August 2016), "The human element", Chemistry World Careers blog, www.chemistryworld.com/careers/what-skills-will-you-need-in-2026/1017241.article.

Gauvreau, D., D. Winickoff and J. Philp (2018), "Engineering biology and the grand challenges: Do we need a new R&D&I model?", *Engineering Biology*, in press.

Glaser, V. (2013), "Overview: Meeting current needs and assessing future opportunities to drive the global bioeconomy", *Industrial Biotechnology*, Vol. 9/5, Mary Ann Liebert, Inc., New Rochelle, United States, pp. 271-274.

Haen, K.M. et al. (2012), "Training the next generation of creative, innovative and adaptive scientists and engineers: The NSF Engineering Research Center for Biorenewable Chemicals (CBiRC) – Research experience for undergraduates", *Education for Chemical Engineers*, Vol. 7, Elsevier, Amsterdam, pp. 230-240.

IBioIC (n.d.), Industrial Biotechnology Information Centre website, www.ibioic.com (accessed 7 December 2017).

Kuldell, N. (2007), "Authentic teaching and learning through synthetic biology", *Journal of Biological Engineering*, Vol. 1/8, Springer, http://dx.doi.org/10.1186/1754-1611-1-8.

Kuntz, M. (2016). "Scientists should oppose the drive of postmodern ideology", *Trends in Biotechnology*, Vol. 34/12, Elsevier, Amsterdam, pp. 943-945.

Lambert, R. (2004), "Dip into the reservoir of innovation", *Cambridge Manufacturing Review*, Spring 2004, University of Cambridge, pp. 2-5.

Liu, D., A. Hoynes-O'Connor and F. Zhang (2013), "Bridging the gap between systems biology and synthetic biology", *Frontiers in Microbiology*, Vol. 4: 211, http://dx.doi.org/10.3389/fmicb.2013.00211.

Mascarelli, A. (2013), "Environmental puzzle solvers", *Nature*, Vol. 494/7438, Nature Publishing Group, London, pp. 507-509.

National Academy of Sciences (2016), "Gene drives on the horizon: Advancing science, navigating uncertainty, and aligning research with public values", National Academies Press, Washington, DC.

National Academy of Sciences (2014a), "The postdoctoral experience revisited", National Academies Press, Washington, DC.

National Academy of Sciences (2014b), "Undergraduate chemistry education. A workshop summary", National Academies Press, Washington, DC.

National Academy of Sciences (2013), "Frontiers in massive data analysis", National Academies Press, Washington, DC.

National Academy of Sciences (1995), "Reshaping the graduate education of scientists and engineers", National Academies Press, Washington, DC.

Nature (2016), "Life after academia", *Nature Editorial*, Vol. 537, Nature Publishing Group, London, pp. 585.

NIBRT (n.d.), National Institute for Bioprocessing Research and Training website, www.nibrt.ie (accessed 7 December 2017).

OECD (2015), *The Innovation Imperative. Contributing to Productivity, Growth and Well-being*, OECD Publishing, Paris, http://dx.doi.org/10.1787/9789264239814-en.

OECD (2005), "Supply and demand chain inhibitors in industrial biotechnology", OECD, Paris, www.oecd.org/officialdocuments/publicdisplaydocumentpdf/?cote=DSTI/STP/BIO(2005)44&docLanguage=En.

Panke, S. (2008), "Synthetic biology – engineering in biotechnology", A report of the Committee on Applied Bioscience, Swiss Academy of Engineering Sciences, Zurich.

Rai, A. and J. Boyle (2007), "Synthetic biology: Caught between property rights, the public domain, and the commons", *PLoS Biology*, Vol. 5/3, Public Library of Science, San Francisco, pp. 0389-0393.

RIKEN (n.d.), RIKEN website, www.riken.jp/en (accessed 7 December 2017).

SynbiCITE (n.d.), SynbiCITE website, www.synbicite.com (accessed 7 December 2017).

Tadmor, B. and B. Tidor (2005), "Interdisciplinary research and education at the biology-engineering-computer science interface: A perspective", *Drug Discovery Today*, Vol. 10, Elsevier, Amsterdam, pp. 1706-1712.

UGA (n.d.), "Master of Biomanufacturing and Bioprocessing", webpage, https://biomanufacturing.uga.edu (accessed 7 December 2017).

Wallman, S. et al. (2013), "Training technicians to support the bioeconomy: Defining the need; designing and implementing innovative solutions", *Industrial Biotechnology*, Vol. 9/5, Mary Ann Liebert, Inc., New Rochelle, United States, pp. 275-281.

Westfall, P.J. and T.S. Gardner (2011), "Industrial fermentation of renewable diesel fuels", *Current Opinion in Biotechnology*, Vol. 22/3, Elsevier, Amsterdam, pp. 344-350.

Woodley, J.M. et al. (2013), "A future perspective on the role of industrial biotechnology for chemicals production", *Chemical Engineering Research and Design*, Vol. 91/10, Elsevier, Amsterdam, pp. 2029-2036.

Glossary

Advanced biofuels. Usually referred to as produced from lignocellulosic sources, these biofuels are produced through the application of advanced conversion processes to crops and novel feedstocks such as algae.

Advanced energy manufacturing tax credit. A tax credit awarded to firms for qualifying investments in renewable and advanced energy projects to support new, expanded or re-equipped domestic manufacturing facilities. For example, the Section 48C tax credit of the American Reinvestment and Recovery Act of 2009 (ARRA) is equal to 30% of the basis of qualifying investments used to manufacture property that will reduce greenhouse gas emissions or air pollutants.

Agenda 2030 for Sustainable Development. The Heads of State and Government and High Representatives met at the United Nations Headquarters in New York from 25 to 27 September 2015 and decided on new global Sustainable Development Goals (SDGs). The 17 SDGs and 169 targets announced are far-reaching in their scale and ambition – this is the "new universal Agenda".

Agricultural residue. Agricultural crop residues are the plant parts, primarily stalks and leaves, not removed from the fields with the primary food or fibre product. Examples include corn stover (stalks, leaves, husks and cobs), wheat straw and rice straw. It can also include other agricultural wastes, including slurry and manure.

Algal lipid biorefinery. Component separation in primary refining results in algal lipids and algal biomass. They generally refer to microalgae, which are often single-celled organisms. Having been grown in a reactor under conditions that allow accumulation of lipids, the lipids are then extracted. The product is an algal oil rich in triglycerides, but also in higher-value materials such as carotenoids and phytosterols. Triglycerides form the basis of biodiesels and are also a potential raw material for the chemical industry.

Aliphatic. Relating to organic compounds whose carbon atoms are linked in open chains, either straight or branched, rather than containing a benzene ring. Alkanes, alkenes and alkynes are aliphatic compounds. They are important molecules in petrochemistry. The alkanes are significant components of liquid transport fuels. The short-chain alkenes, such as ethylene and propylene, are at the heart of the petrochemicals industry.

Anaerobic digestion. Degradation of organic matter by microbes that produces a gas comprised mostly of methane and carbon dioxide in the absence of oxygen.

Aromatic. The term was coined simply because many of the aromatic compounds have a sweet or pleasant odour. Aromatic hydrocarbons contain six carbon atoms in a ring structure known as a benzene ring, after the simplest one, benzene. Aromatic compounds have many uses. The aromatic ring can be found in rubbers, lubricants, dyes, detergents, drugs, explosives and pesticides, among others. Apart from their widespread utility, many are toxic to very toxic, and some are known to cause cancer. Bio-based versions are difficult to manufacture.

B20. A mixture of 20% biodiesel and 80% petroleum diesel based on volume.

Bagasse. Residue remaining after extracting a sugar-containing juice from plants like sugar cane.

Barrel (of oil). A liquid measure equal to 42 US gallons (35 Imperial gallons or 159 litres); about 7.2 barrels are equivalent to 1 tonne of oil (metric).

Benzene. A six-sided structure with three alternating double bonds. It is a known carcinogen that is an aromatic component of petrol.

Bio-based chemicals tax credit. The US Renewable Chemicals Act of 2015 would create a 15 cent-per-pound production tax credit for eligible renewable chemicals manufactured from biomass feedstock. Alternatively, the bill would allow producers to take a 30% investment tax credit for qualified investments for new renewable chemical production facilities in lieu of the production tax credit. The Qualifying Renewable Chemical Production Tax Credit Act of 2012 would provide renewable chemical and bio-based products access to tax credits that are available to other industries.

Bio-based content. The amount of bio-based carbon in the material or product as a percent of weight (mass) of the total organic carbon in the material. This is an important indicator of renewability, but not necessarily of sustainability. Bio-based products need not be composed entirely of bio-based carbon. The emerging bio-based manufacturing industry produces large quantities of products that contain mixtures of bio-based and fossil-derived materials, e.g. first-generation bio-polyethylene terephthalate (bio-PET, the material commonly used to make drink bottles).

Bio-based product. A product made partially or entirely from substances derived from living matter. It may include common materials such as wood and leather, but typically means modern materials that have undergone more extensive processing. Bioproducts or bio-based products include materials, chemicals and energy derived from renewable biological resources. Bio-based materials are often, but not necessarily, biodegradable. The term is typically applied only to materials containing carbon.

Biocoal. This is a solid fuel made from biomass by heating it in an inert atmosphere. The result is either charcoal, or if the process temperature is mild, a product called "torrefied wood". Charcoal and torrefied wood can be called by the common name biocoal. Compared to untreated biomass, biocoal has several advantages. It has high energy content, uniform properties and low moisture content.

Biodegradable. Capable of being decomposed by biological agents, especially bacteria and fungi. Biodegradable has proven a controversial term as it does not necessarily mean biodegradation to its mineral components, despite the implications of marketing. Partial biodegradation can, in fact, result in a stable intermediate compound that is more toxic than the original molecule. Mineralisation means the ultimate conversion of the material or compound to its mineral components (CO_2 and H_2O under aerobic conditions). Biodegradability usually refers to the testing regime under which a compound or material is judged to be biodegradable or not.

Biodiesel. Biodiesel is an alternative to fossil diesel fuel in transport. It is similar in composition, but can be produced from straight vegetable oil, animal oil/fats, tallow and waste cooking oil. A biodiesel of the future will be derived from algae. Biodiesel can be used alone, or blended with petro-diesel in any proportions. Indeed, many renewable energy policies depend on blending. Biodiesel blends can also be used as heating oil.

Biodiversity. The variety of all life on Earth, including all species of animals and plants, and the natural systems that support them. There are ongoing studies on the links between climate change and biodiversity from at least two perspectives: impacts of climate change and climate policy on biodiversity and ecosystem services, and the role of biodiversity and ecosystem services in climate change mitigation and adaptation.

Biocatalyst. Usually refers to enzymes and microbes, but it can include other catalysts that are living or that were extracted from living organisms, such as plant or animal tissue cultures, algae, fungi or other whole organisms.

Biocrude. A crude oil similar to petroleum that can be produced from biomass under high pressure and temperature.

Biodiesel. Conventionally defined as a biofuel produced through transesterification, a process in which organically-derived oils are combined with alcohol (ethanol or methanol) in the presence of a catalyst to form ethyl or methyl ester. The biomass-derived ethyl or methyl esters can be blended with conventional diesel fuel or used as a neat fuel (100% biodiesel). Biodiesel can be made from soybean or rapeseed oils, animal fats, waste vegetable oils or microalgae oils. (Note: Biodiesel can in certain circumstances include ethanol-blended diesel. This is an evolving definition.)

Bioeconomy. Since a landmark OECD publication, bioeconomy definitions have varied, and as yet there is no consensus. From a broad economic perspective as envisaged by the OECD, the bioeconomy refers to the set of economic activities relating to the invention, development, production and use of biological products and processes. However, it has come to include agriculture, forestry, pulp and paper, and other sectors. Some countries include health in the definition. Therefore, estimates of the size of the bioeconomy vary enormously.

Bioenergy. Energy generated by combusting solid, liquid or gas fuels made from biomass feedstocks, which may or may not have undergone some form of conversion process. Organic matter may either be used directly as a fuel processed into liquids and gases, or be a residual of processing and conversion.

Bioethanol. Bioethanol is the principal fuel used as a petrol substitute for road transport vehicles, generally produced from crops such as sugar cane, corn and wheat. It can be made from virtually any biomass source (grass, wood, biodegradable elements of municipal solid waste), but the technologies for doing so are still under development.

Biofuel. A fuel produced from biomass feedstocks. Strictly speaking, fossil fuels fit this definition, but biofuels are distinguished from fossil fuels in that they are produced from renewable biomass, i.e. crops that can be harvested for refinement to biofuels, and replanted and reharvested on a continuing basis.

Biofuel intermediate. A biomass-based feedstock that serves as a petroleum replacement in downstream refining, (i.e. sugars, intermediate chemical building blocks, bio-oils and gaseous mixtures). Algal biofuel intermediates include extracted lipids, lipid-extracted biomass or bio-oil resulting from hydrothermal liquefaction.

Biofuels sustainability criteria. In the European Union, to be considered sustainable (and therefore qualify for government support), biofuels must achieve greenhouse gas savings of at least 35% in comparison to fossil fuels. This savings requirement rises to 50% in 2017. In 2018, it rises again to 60%, but only for new production plants. All life cycle emissions are taken into account when calculating greenhouse gas savings. This includes emissions from cultivation, processing and transport. In addition, biofuels cannot be grown in areas converted from land with previously high carbon stock such as wetlands or forests; and they cannot be produced from raw materials obtained from land with high biodiversity such as primary forests or highly biodiverse grasslands.

Biogas. Biogas typically refers to a mixture of different gases produced by the breakdown of organic matter in the absence of oxygen. Biogas can be produced from raw materials such as agricultural waste, manure, municipal waste, plant material, sewage, green waste or food waste. It is typically depicted as a mixture of methane and carbon dioxide with traces of many other gases possible in some sources e.g. landfill gas.

Biogas biorefining. Here there is no separate component separation in primary refining. The organic materials present in the feedstock are anaerobically decomposed in a complex microbiological process to biogas, comprised mainly of methane and CO_2. The gas is flammable and can be burned to produce heat and electricity. In wastewater treatment plants, this is traditionally called anaerobic digestion, and can in many cases meet the electricity needs of the entire plant.

Biomass. This is the biological raw material used to make fuels or other bio-based products. It includes solid biomass such as wood, plant and animal products, gases and liquids derived from biomass, and the biodegradable components of industrial and municipal wastes. Processing and conversion derivatives of organic matter are also biomass.

Biomass potential. Biomass potential refers to the amount of biomass that can be grown. In the modern policy setting, this will refer to the sustainable biomass potential. Several biomass potential studies have been done in the last few decades. Their approaches have been very different and their results difficult to compare and interpret. They can be done at local, regional, national levels or above. Without standardised criteria for measuring biomass potential, future estimates will continue to be uncomparable and variable.

Biomass sustainability. The meaning of biomass sustainability depends on what is meant by sustainability. The latter has come to be associated with safeguarding the future by not taking out more from the planet than necessary. This responds to the habit of rapid population growth and development being accompanied by a "throw-way" mentality when resources are finite. Therefore, it calls for using renewable resources and higher levels of recycling.

Bioplastic. There is no universally agreed definition of a bioplastic. Bioplastics were first introduced as biodegradable plastics for use in simple packaging applications. The bio-based plastics that are now increasing in the market are not necessarily biodegradable, but they contain carbon that is partially or entirely derived from renewable biomass.

Bio-principled cities. The integration of biological principles into urban planning and city life. It calls for higher levels of self-sufficiency of cities and more recycling to reduce waste and close energy and material loops.

Blending tax credit. Biofuels blenders are eligible for an income tax credit per litre or gallon. For example, under the US Biodiesel Production and Blending Tax Credit, qualified biodiesel producers or blenders are eligible for an income tax credit of USD 1.00 per gallon of pure biodiesel (B100) or renewable diesel produced or used in the blending process. For the purpose of this credit, biodiesel must meet ASTM specification D6751, and renewable diesel is defined as a "renewable, biodegradable, non-ester combustible liquid derived from biomass resources that meets ASTM specification D975". The blending tax credit has been criticised for being accessible to foreign biofuels; a producer's credit would support domestic production more. On the other hand, proponents of the blender's credit say that a producer-only credit increases profits for a limited number of producers, while reducing the overall availability of fuels.

Business ecosystem/ecology. A system-level view of the relationships and interdependencies evident in organisations, markets or industries, including their components, actors, resources and stakeholders. Inspired from nature, this is a similar approach to understanding biological ecosystems in their fullest sense.

By-products. These occur as a result of primary- and/or secondary refining, and are used to supply process energy or, where applicable and in compliance with statutory requirements, are further processed into food or feed.

Cap and trade. A cap is placed on the total amount of allowable emissions, it is distributed among the total number of polluters and a marketplace is created where owners of the permits can trade with each other. The intention is to incentivise a reduction in emissions and penalise those who fail to comply.

Capital cost. The total investment needed to complete a project and bring it to an operable status. The cost of construction of a new plant. The expenditures for the purchase or acquisition of existing facilities.

Carbon capture and storage (CCS). This technology involves capturing CO_2, transporting it and storing it in secure spaces such as geological formations, including old oil and gas fields, and aquifers under the seabed.

Carbon dioxide equivalents (CO_2e). The internationally recognised way of expressing the amount of global warming of a particular greenhouse gas in terms of the amount of CO_2 required to achieve the same warming effect over 100 years.

Carbon footprint. The total emissions of greenhouse gases (in carbon equivalents) from whichever source is being measured – be it at an individual, organisation or product level.

Carbon-neutral. Carbon neutrality makes or results in no net release of carbon dioxide into the atmosphere, especially as a result of carbon offsetting.

Carbon offsetting. The process of reducing greenhouse gas emissions by purchasing credits from others through emissions reductions projects or carbon trading schemes. The term often refers to voluntary acts, arranged by a commercial carbon offset provider.

Carbon price and carbon tax. A carbon price is the amount that must be paid for the right to emit one tonne of CO_2 into the atmosphere. Carbon pricing usually takes the form either of a carbon tax or a requirement to purchase permits to emit, generally known as cap and trade. Carbon pricing has proven extremely controversial politically. As a consequence of not being priced, there is no market mechanism responsive to the costs of CO_2 emitted. Classically, emissions should be charged at a price equal to the monetary value of the damage caused by the emissions. This should result in the economically optimal (efficient) amount of CO_2 emissions. However, the price of the damage has remained elusive.

Carbon trading. Any trading system designed to offset carbon emissions from one activity (such as burning fossil fuels in manufacturing, driving or flying) with another (such as installing more efficient technologies, planting carbon-reducing plants or establishing contracts with others not to partake in carbon-releasing activities). When activities that reduce or capture carbon are paired successfully with those that produce it, these are said to be carbon neutral or climate neutral.

Cascading use. This usually refers to the cascading use of biomass. The theory goes that the highest value products, generally in the lowest volumes, are extracted from biomass first. Then the same biomass cascades towards the lowest value products, often the highest-volume materials. Recycling is applied as often as possible before the biomass and its products reach the end of their life, perhaps by burning to generate electricity. In this manner, the maximum value is theoretically extracted from the original biomass. It has been estimated that cascading can lead to an almost 30% reduction in European greenhouse gas emissions by 2030 compared with 2010.

Certification schemes and labels. In certification schemes, independent organisations test materials or products. If results are satisfactory, the organisations issue a certificate stating that the material or product meets the requirements (prescriptions) of a particular standard. Certification of bio-based products informs users about the nature of the material or product. Certification is often accompanied by a label that may be placed on certified materials and products.

Circular economy. This is strongly associated with recycling. It refers to closed loops of energy and material use in that the residues and by-products of one process can be used in another. The ultimate goal of a circular economy would be "zero waste". A strong relationship to cascading use will be evident.

Clean production. Manufacturing processes designed to minimise environmental impact by using the minimum amount of energy and raw materials possible and producing limited waste or emissions.

Clear cutting. A process where all trees in a selected area are felled in a logging operation. This can be extremely destructive to the environment, while being the most cost-effective means known to harvest high yields of timber rapidly.

Climate change. Climate change has come to be associated with global warming as a result of human activities since the Industrial Revolution, but it is also caused by factors such as oceanic processes, variations in solar radiation, plate tectonics and volcanic eruptions. The term is now almost universally used to describe impacts resulting from human activity.

Combined Heat and Power (CHP). A system in which the heat associated with electricity generation is used for space heating or process heat. It considerably increases the overall efficiency of the fuel used in the process. Energy generated by the incineration of waste at local combined heat and power facilities can support district heating schemes.

Composting. In the context of sustainability this refers to a regulated industrial-scale process for converting decomposable organic materials into useful stable products through biochemical processes. Industrial-scale composting through in-vessel composting, aerated static pile composting and anaerobic digestion is now used in most Western countries and is often legally mandated. Composting is one of the very few ways to revitalise soil in which the phosphorus is depleted.

Conventional biofuels. These are transport biofuels typically derived from crops and waste using current conversion processes. Examples include bioethanol from sugar cane and biodiesel from oilseed rape and used cooking oil. These are also known as first-generation biofuels.

CO_2 economy. A CO_2 economy encompasses both carbon capture and storage (CCS) and carbon capture and use (CCU). For example, hot, pressurised CO_2 can be used not only for generating power, but also for higher-value, carbon-negative products, such as synthetic gasoline and diesel fuel. This ushers in an era of possibilities: clean, reliable baseload energy; a cost-effective means to capture greenhouse gases; and the affiliated production of potent carbon-negative, fossil fuel substitutes. But the CO_2 economy can also be associated with a low-carbon economy (due to the renewable and recyclable elements of the theory).

Cradle-to-cradle. A design protocol that advocates the elimination of waste by recycling a material or product into a new or similar product at the end of its intended life, rather than disposing of it.

Cradle-to-gate. An assessment of a partial product life cycle from manufacture ("cradle") to the factory gate i.e. before it is transported to the user or consumer. The use phase and disposal phase of the product are usually omitted. Cradle-to-gate assessments are sometimes the basis for environmental product declarations. They are of greatest use to manufacturers, who cannot foresee what customers will do with their products.

Cradle-to-grave. A manufacturing model that describes the process of disposing of a material or product via a recognised route, such as landfill or incineration, at the end of its presumed useful life.

Dedicated energy crops. These are crops grown to be used for energy generation. Examples include fast-growing trees (such as short rotation coppice willow) and grasses with a high lignocellulosic content (such as *Miscanthus*).

Deforestation. This is the clearing of the planet's forests on a massive scale, often resulting in damage to the quality of the land, and reducing the capacity of the planet to absorb CO_2. An estimated 13 million hectares of forests were lost each year between 2000 and 2010 due to deforestation.

Direct land-use change (DLUC). The conversion of land from one use to another, e.g. from unmanaged forest to cropland, or from one crop type to another. The tillage of unmanaged land exposes large amounts of soil organic carbon to the atmosphere and produces large amounts of CO_2. It can take a long time to pay back this CO_2 debt.

Drop-in fuel. A substitute for conventional fuel that is completely interchangeable and compatible with conventional fuel. A drop-in fuel does not require adaptation of the engine, fuel system or the fuel distribution network and can be used "as is" in available engines in pure form and/or blended in any amount with other fuels.

Eco-label. An environmental label or declaration that provides information about a product or service in terms of its overall environmental character, a specific environmental aspect or number of environmental aspects. The information can be used to influence or inform purchasing decisions. Eco-labels may take the form of a statement, symbol or graphic and be found, in part, on products or packaging and in product literature or advertising.

E-10. A mixture of 10% ethanol and 90% petrol based on volume. In the United States, E-10 is the most commonly found mixture of ethanol and petrol.

E-85. Typically refers to an ethanol fuel blend of 85% denatured ethanol fuel and 15% petrol or other hydrocarbon by volume.

Emission. The release of any gas, particle or vapour into the environment from a commercial, industrial or residential source, including smokestacks, chimneys and motor vehicles.

Emissions trading. This refers to the trading of permits that allow emissions of set amounts of greenhouse gases. It is therefore a market mechanism for controlling pollution. By creating tradable pollution permits, it attempts to add the profit motive as an incentive for good performance, unlike traditional environmental regulation based solely on the threat of penalties.

End-of-life options. This refers to the step in the life of a chemical or material after its primary intended purpose has been fulfilled. The chemical or material may then be used for another purpose. Often it becomes a waste product. Here the options typically are discarding to landfill, incineration (with or without energy capture and electricity production), composting (for biodegradable wastes) and recycling for further use.

Energy crops. Crops grown specifically for their fuel value. These crops may include food crops such as corn and sugarcane, and non-food crops such as poplar trees and switchgrass.

Energy density. In terms of fuels, energy density means the amount of energy stored in a given liquid, per unit volume (litre or gallon, normally). Ethanol has a higher energy density than methanol, meaning a vehicle can be driven farther on a litre of ethanol than a litre of methanol. Biomass is also described as being of low energy density, making it inefficient to transport over long distances.

Energy intensity. There are two meanings in common use. First, it can be a measure of the energy efficiency of a national economy, calculated as units of energy per unit of gross domestic product. Second, it can be the entire amount of energy required to produce a product as a ratio of that product.

Energy recovery. This is obtaining energy from waste. It is accomplished through a variety of processes, and is also known as "waste-to-energy". Traditionally, this meant burning waste products, but now gasification and anaerobic digestion are also playing a role.

Energy security. At its core, energy security is the concept of physical security (avoiding involuntary interruptions of supply). It can include elements of price security (e.g. avoidance of excessive price volatility). In the context of sustainability and affordability and as a result of volatility of fossil fuel prices, many countries seek to improve energy security through diversification of supply, e.g. biofuel production, offshore and onshore wind energy, solar power.

Enzymatic hydrolysis. Use of an enzyme to promote the conversion, by reaction with water, of a complex substance into two or more smaller molecules.

Enzyme. One of various proteins that act as catalysts for a single reaction, converting a specific set of reactants into specific products.

Eutrophication. The process by which a body of water becomes enriched in dissolved nutrients (such as phosphates) that stimulate the growth of aquatic plant life usually resulting in the depletion of dissolved oxygen that can lead to fish kills and loss of other life forms.

Externality. A cost or benefit not accounted for in the price of goods or services. Often refers to the cost of pollution and other environmental impacts.

Feed-in tariff. A feed-in tariff (FIT), or advanced renewable tariff or renewable energy payments is a policy mechanism designed to accelerate investment in renewable energy technologies. It achieves this by offering long-term contracts to renewable energy producers, typically based on the cost of generation of each technology. Producers may be generating companies of any size, all the way down to the individual home. It is effectively a scheme that pays people for creating their own green or renewable electricity. For governments, FITs help meet national production targets on renewable energy and therefore for greenhouse gas emissions reductions.

Feedstocks. Any material converted to another form or product, or the starting material for a process. It includes crops or products that can be used to produce bioenergy or bio-based products. Examples are wood, switchgrass, waste paper, agricultural residues, corn and soybeans.

Fermentation. The use of microorganisms (e.g. yeasts, bacteria) to break down organic substances to create other organic substances. The classic fermentation is the anaerobic conversion of sugars into ethanol by the yeast *Saccharomyces cerevisiae*. In strict scientific terms, fermentation is an anaerobic process, but in practice the term is also used for aerobic biological conversion processes.

Fischer-Tropsch. The Fischer-Tropsch process is a set of chemical reactions that converts a mixture of carbon monoxide gas and hydrogen gas into liquid hydrocarbons (fossil fuels like gasoline or kerosene).

Flex-fuel vehicle (FFV). A vehicle that operates with more than one fuel or fuel blend. For example, the Ford Model T, produced from 1908 to 1927, was fitted with a carburettor with adjustable jetting, allowing use of petrol or ethanol, or a combination of both. FFVs have been designed for modern use with petrol or ethanol, and blends thereof, in many countries as a means of reducing emissions, most notably in Brazil and Sweden.

Food security. Food security refers to the availability of food. In the bioenergy context, it relates to the food vs. fuel debate about diverting farmland from food crops to bioenergy feedstocks and the perception that this leads to higher food prices and decreased food supply worldwide. The origins of the recent food price spikes, however, are complex and include other issues such as the price of fossil fuels.

Forestry and forest residues. These forest sector by-products include residues from thinning and logging (e.g. treetops, limbs) and secondary residues such as sawdust and bark from wood processing. Forestry and forest residues also include dead wood from natural disturbances, such as fires, biomass grown in forests that is not required for timber production, and biomass from dedicated plantations, e.g. short- and long-rotation forestry.

Fossil fuel. Coal, oil and gas are called fossil fuels because they were formed from the fossil remains of plants and animals millions of years ago. As fuels, they offer high energy density, but making use of that energy requires burning the fuel and the oxidation of the carbon to CO_2 and H_2O. Unless they are captured and stored, these combustion products are released and return carbon sequestered millions of years ago to the atmosphere.

Fossil fuel subsidies. These are any government actions that lower the cost of fossil fuel energy, that raise the price received by energy producers or lower the price paid by energy consumers. There are a lot of activities under this simple definition – tax breaks and giveaways, but also loans at favourable rates, price controls, purchase requirements and more. The global value of these subsidies is vast, by most calculations of the order of half a trillion USD per annum. They therefore have the global effect of distorting the fossil fuel markets. For example, the price of petrol in a Western European country can easily be more than one hundred times the price in Venezuela at a given time.

Fuel cell. A device that converts the energy of a fuel directly to electricity and heat, without combustion.

Gasification. The process that converts organic or fossil fuel-based carbonaceous materials into CO, H_2 and CO_2, and possibly hydrocarbons such as CH_4. This is achieved by reacting the material at high temperatures without combustion, with a controlled amount of oxygen and/or steam. It can therefore be viewed as a "partial oxidation" process.

Genetically modified. An organism is genetically modified if it contains genetic material that has been artificially altered to produce a desired characteristic. The term has become extremely political and societally divisive, leading to multiple interpretations of its meaning. The greatest controversy is with food. Genetically modified (GM) foods are foods derived from organisms whose genetic material (DNA) has been modified in a way that does not occur naturally, e.g. through the introduction of a gene from a different organism.

Genome. A genome is the complete set of DNA, including all of the genes, of an organism. Each genome contains all of the information needed to build and maintain that organism. In humans, a copy of the entire genome – more than 3 billion DNA base pairs – is contained in all cells that have a nucleus.

Genomics. Now a rather broad term, it is the branch of molecular biology concerned with the structure, function, evolution and mapping of genomes.

Global warming. The gradual increase in the overall temperature of the Earth's atmosphere generally attributed to the greenhouse effect caused by increased levels of carbon dioxide, CFCs and other pollutants. By overwhelming scientific consensus, global warming is caused by human activity.

Green bank. A green bank is a public or quasi-public financing institution that provides low-cost, long-term financing support to clean, low-carbon projects by leveraging public funds using various financial mechanisms to attract private investment. In this way, each public dollar supports multiple dollars of private investment. An early example is the UK Green Investment Bank. Typical projects could include offshore and onshore renewable energy, offshore wind, solar, energy efficiency, waste and bioenergy.

Green chemistry. The design of chemical products and processes that reduce or eliminate the use and generation of hazardous substances. Hazardous now has come to include greenhouse gases, in particular CO_2. There is now an emphasis on the link between green chemistry and climate change/global warming.

Green engineering. Engineering with environmentally conscious attitudes, values and principles, combined with science, technology and engineering practice, all directed towards improving local and global environmental quality.

Green growth. This term describes a path of economic growth that uses natural resources in a manner that is sustainable. It is used globally to provide an alternative concept to typical industrial economic growth. A green growth strategy would bring together the three pillars of sustainable development (economic, environmental and social), and incorporate technological and development aspects into a comprehensive framework.

Greenhouse effect. The greenhouse effect is the process by which radiation from the atmosphere of the planet warms its surface to a temperature above what it would be without its atmosphere.

Greenhouse gas (GHG). Any atmospheric gas (either natural or of human origin) that absorbs thermal (infrared) radiation emitted by the Earth's surface. This traps heat in the atmosphere and keeps the surface at a warmer temperature than would otherwise occur. The primary greenhouse gases in the Earth's atmosphere are water vapour, carbon dioxide, methane, nitrous oxide and ozone.

Greenwashing. A term merging the concepts of "green" (environmentally sound) and "whitewashing" (to conceal or gloss over wrongdoing). Greenwashing is any form of marketing or public relations that links a corporate, political, religious or non-profit organisation to a positive association with environmental issues for an unsustainable product, service or practice.

Gross domestic product (GDP). A measure of economic production (and often standard of living) of a country. GDP calculates a nation's total economic output of products and services. The GDP is problematic as a sustainability indicator because it considers the amount of money spent in a country in isolation, assuming more money spent means a healthier economy.

Gross national product (GNP). The total value of newly produced products and services produced in a year by a country's companies (including profits from capital held abroad). Transactions in existing goods, such as second-hand cars, are not included, as these do not involve the production of new goods.

Hydrocarbon. A compound of hydrogen and carbon, such as any of those that are the chief components of petroleum and natural gas.

Hydrogenation. This means to treat with hydrogen. It is a chemical reaction between molecular hydrogen (H_2) and another compound or element, usually in the presence of a catalyst. The process is commonly employed to reduce or saturate organic compounds. Hydrogenation is becoming an important tool in the efficient conversion of biomass to value-added products.

Hydrogen economy. The term refers to the vision of using hydrogen as a low-carbon energy source – replacing, for example, petrol as a transport fuel or natural gas as a heating fuel. However, its "green" credentials are questioned if the hydrogen is generated by steam reformation of hydrocarbons. Other means of generation, such as water electrolysis, require large energy inputs. Bio-based hydrogen to date suffers from poor production and therefore applications at large scale are still evanescent.

Hydrolysis. In relation to biorefining, the hydrolysis of biomass is usually taken to mean the hydrolysis of cellulose present in biomass to produce sugars and other organic compounds that can be subsequently fermented. In chemistry, it literally means the chemical breakdown of a compound due to reaction with water.

Hydrothermal. Hydrothermal processing of biomass is similar to torrefaction, but uses milder treatment conditions. Hydrothermally processed biomass is commonly referred to as "biocoal".

Incineration. Incineration of waste materials converts the waste into ash, flue gas and heat. The flue gases should be cleaned of gaseous and particulate pollutants before they are dispersed into the atmosphere. In some cases, the heat generated by incineration can be used to generate electric power. Incinerators reduce the solid mass of the waste by 80-85% and the volume by 95-96%. Incineration does not completely replace landfilling, but significantly reduces the volume to be disposed of.

Indirect land-use change (ILUC). Indirect land-use change occurs when land used for an existing activity (e.g. food or timber production) is converted to grow a bioenergy feedstock or when a food crop is used for bioenergy (e.g. diversion of maize for ethanol production). It is assumed that food production is essential and that the lost food production will be diverted elsewhere.

Industrial biotechnology. Industrial or white biotechnology uses enzymes and microorganisms to make bio-based products in sectors such as chemicals, food and feed, detergents, paper and pulp, textiles and bioenergy (such as biofuels or biogas).

Industrial ecology. A field of study and practice that focuses on how industry can be developed or restructured to reduce environmental burdens throughout the product life cycle (extraction, production, use and disposal). In applying this perspective, companies seek to shift industrial processes from open-loop systems that produce waste materials to closed loop systems where wastes become inputs for new processes. Perhaps the best-known example is Kalundborg, Denmark.

Industrial metabolism. The total use of materials and energy throughout an entire industrial process, such as manufacturing. This includes the source, transportation, use, reuse, recycling and disposal of all industrial nutrients (materials), as well as the energy needed at each step.

Irrigation. Irrigation becomes an issue of sustainability because about 70% of fresh water use is for agriculture. Much of this is used for irrigation. Irrigation is the artificial application of water to the land or soil. It is used to assist in the growing of agricultural crops, maintenance of landscapes and revegetation of disturbed soils in dry areas and during periods of inadequate rainfall. Methods vary greatly in their efficiency. One of the drives in plant genomics is to produce crop varieties that are heat- and drought-tolerant to reduce use of water in agriculture, thereby improving water security.

Integrated biorefineries. In the integrated biorefinery model, multiple products are made at the same facility or complex – biofuels, bio-based materials and bioenergy. Often the economics of bio-based chemicals production will be superior to those of biofuels production (this is also the case in petro-production). Integration is widely accepted as the most sustainable form of biorefining for the future.

Kyoto Protocol. Adopted in 1997 as a protocol to the United Nations Framework Convention on Climate Change (UNFCCC), the Kyoto Protocol placed a legally binding commitment on participating countries to reduce their greenhouse gas emissions by 5% relative to 1990 levels over the period from 1998 to 2012. Gases covered by Kyoto Protocol are carbon dioxide (CO_2), methane (CH_4), nitrous oxide (N_2O), sulphur hexafluoride (SF_6), hydroflurocarbons (HFCs) and perfluorocarbons (PFCs).

Landfill. In the sustainability context, landfill refers to ground filled in with waste materials; in other contexts, it can refer to rocks. Landfill is the most common organised waste disposal technology. As waste is buried in landfill, the site rapidly becomes anaerobic, with the result that materials that biodegrade produce methane, which can be captured and burned for heat or electricity production. If the methane production is not controlled, it contributes to GHG emissions. Decades of concern over the declining number of suitable sites for landfilling has led to legislation to limit materials that are landfilled.

Life cycle analysis (LCA). This is a numerical technique to assess environmental impacts associated with all the stages of the life of a product, typically from cradle-to-grave (i.e. from raw material extraction through materials processing, manufacture, distribution, use, repair and maintenance, and disposal or recycling. There are several variants of cradle-to-grave in common use.

- Cradle-to-factory gate assesses the product life cycle from resource extraction (cradle) to the factory gate i.e. before it is transported to the consumer. This variant is often favoured by the manufacturer, which does not control the consumer's use and disposal of the product.
- Cradle-to-cradle (also known as open-loop production) is a cradle-to-grave assessment for which recycling is the end-of-life option.
- Well-to-wheel is a specific LCA used for transport fuels and vehicles.

Lifecycle emissions. The emissions generated by a product, system or service over its lifetime.

Lignin. An amorphous polymer related to cellulose that, together with cellulose, forms the cell walls of woody plants and acts as the bonding agent between cells.

Lignocellulosic biorefining. In dry biomass-based lignocellulosic biorefining, the component separation in the primary refining produces the lignocellulosic components cellulose, hemicelluloses and lignin, extremely common components of plant and woody materials. The feedstocks are various e.g. straw, forestry trimmings, other agricultural residues and dedicated energy crops.

Lignocellulosic feedstock. Woody feedstocks with significant cellulose and hemicellulose content. Advanced conversion processes are required to break down the cellulose and hemicellulose for conversion to liquid biofuels or bio-based products. The biorefining of these feedstocks involves a high cost.

Loan guarantee. A loan guarantee in finance is a promise by one party (the guarantor) to assume the debt obligation of a borrower if that borrower defaults. A guarantee can be limited or unlimited, making the guarantor liable for only a portion or all of the debt. This has become a major finance instrument for the construction of high-risk and flagship (first-of-kind) biorefineries.

Managed forests. In a managed forest, trees are replanted as they are felled. Wood products that come from well-managed forests offer the most benefits in terms of combating climate change. Well-managed woodlands also generally store more carbon than stands that are not harvested.

Market failure. A market's inability to create maximum efficiency by not properly providing goods or services to consumers, not efficiently organising production or not serving the public interest. The term does not refer to the collapse or demise of a market.

Materials audit. The process of investigating the costs and effects of materials used in manufacturing in order to determine more efficient, less costly, less toxic (or dangerous) and more sustainable options.

Metabolic engineering. The use of genetic engineering to modify the metabolism of an organism. It can involve the optimisation of existing biochemical pathways or the introduction of pathway components, most commonly in bacteria, yeast or plants, with the goal of high-yield production of specific molecules for medicine or biotechnology. It has many applications in bio-production of chemicals, plastics, textiles and other materials.

Metabolomics. The comprehensive analysis of all low-molecular-weight primary and secondary metabolites present in and around cells growing under defined physiological conditions. It is emerging as a rapidly developing field of research with the promise to speed up the functional analysis of genes of unknown function. The metabolome is the final downstream product of gene transcription. Additionally, as the furthest downstream product, the metabolome is closest to the phenotype of the biological system being studied.

Metagenomics. The application of modern genomics technologies to microbial communities in their natural environments, bypassing the need for culturing. The vast majority of bacterial life, for example, remains unculturable using available methods. For almost the entire history of microbiology as a discipline, perhaps 90-99% of the diversity of bacteria has been a complete mystery.

Municipal solid waste (MSW). Any organic matter, including sewage, industrial and commercial wastes, from municipal waste collection systems. Municipal waste does not include agricultural and wood wastes or residues.

Net present value (NPV). The value, in the present, of an investment or financial transaction that will pay off in the future, minus the cost of the investment up until the time of that pay-off. NPV represents the profit or loss, in present worth, of future transactions so they are comparable against other investments. Large, time-consuming construction projects are managed on this basis e.g. developing an oilfield.

Non-renewable energy. Energy derived from sources that cannot be replenished in a short period of time relative to a human life span. Non-renewable sources of energy are typically divided into two types: fossil fuels and nuclear fuels. Fossil fuels include oil, natural gas and coal. Nuclear involves uranium.

Open-loop recycling. The conversion of material from one or more products into a new product, involving a change in the inherent properties of the material itself e.g. recycling plastic bottles into plastic drainage pipes. This is also referred to as downcycling or reprocessing.

Operating cost. The expenses related to the operation of a business, or to the operation of a device, component, piece of equipment or facility. They are the cost of resources used by an organisation just to maintain its existence.

Organic compound. Compound containing carbon chemically bound to hydrogen. Often contains other elements (particularly O, N, halogens or S).

Oxygenate. A compound that contains oxygen in its molecular structure. Ethanol and biodiesel act as oxygenates when they are blended with conventional fuels. Oxygenated fuel improves combustion efficiency and reduces tailpipe emissions of CO.

Paris Agreement, 2015. The Paris Agreement is an agreement within the framework of the United Nations Framework Convention on Climate Change (UNFCCC) governing greenhouse gas emissions mitigation, adaptation and finance from 2020. The agreement was negotiated during the 21st Conference of the Parties of the UNFCCC in Paris and adopted by consensus on 12 December 2015, but has not entered into force. In the 12-page Agreement, the members promised to reduce their carbon output "as soon as possible" and to do their best to keep global warming "to well below 2 degrees C".

Peak oil. A controversial concept, peak oil is the hypothetical point in time when the global production of oil reaches its maximum rate, after which production will gradually decline (some models envisage precipitous decline). It has political dimensions that pertain to sustainability as it can be used as an argument for greater deployment of renewable energy and materials production.

Precautionary principle. An approach to determining whether a given process or policy should be pursued or continued based on an analysis of the social, economic or environmental risks associated with that activity. Not all risks are known when a new practice is introduced or a current one is re-examined. The ethical approach in light of implied or expected (but not confirmed) negative impacts is to stop such practices as a precaution until more is known about the impacts.

Pollution prevention. Any activity to reduce or eliminate any number of pollution types or quantities from personal, corporate or governmental activities. These activities seek to create more efficient procedures or practices that reduce pollution or use them in the manufacturing process of some other activity. Many countries have integrated pollution prevention and control into policy. The European Union has the Integrated Pollution Prevention and Control (IPPC) Directive that requires industrial and agricultural activities with a high pollution potential to have a permit. This permit can only be issued if certain environmental conditions are met so that the companies themselves bear responsibility for preventing and reducing any pollution they may cause.

Polymer. A large molecule made by linking smaller molecules (monomers) together.

Polysaccharide. A carbohydrate consisting of a large number of linked simple sugar, or monosaccharide, units. Examples of polysaccharides are cellulose and starch.

Primary refining. This involves the separation of biomass components into intermediates (e.g. cellulose, starch, sugar, vegetable oil, lignin, plant fibres, biogas, synthesis gas), and usually also includes the pre-treatment and conditioning of the biomass. While component separation takes place at the biorefinery, one or more pre-treatment/conditioning processes can also be decentralised as needed.

Production (biofuels) mandate. Under the US Renewable Fuel Standard (RFS), first established in 2005, Congress mandated biofuels use for the United States. The US Energy Independence and Security Act (EISA, 2007) superseded and greatly expanded the biofuels mandate to 36 billion gallons by 2022. This was a substantial biofuels policy as it set minimum usage requirements for various road transport biofuels to guarantee them a market irrespective of their cost.

Proteomics. Proteomics is the large-scale study of proteins, particularly their structures and functions, and the proteome is the entire set of proteins produced or modified by an organism or system. Proteomics lagged behind genomics for a long time due to technical difficulties. However, this has progressed radically in recent years and is now on par with most genomic technologies in throughput and comprehensiveness.

Protein. A long chain of amino acids, folded into a more or less compact structure.

Public-private partnership (PPP). Widely used in science and technology policy, public-private partnership models of different varieties exist. The general idea is that public (taxpayers') money pump-primes greater financial support from the private sector. This signals policy stability from governments and de-risks private investments when the private sector is unwilling to shoulder the risk alone. PPPs are well- suited to the high risk associated with biorefinery projects, especially the integrated biorefineries of the future.

Public procurement. This can be a powerful market-making measure. It involves mass purchasing of commodities by the public sector. Obvious examples might be military procurement of biodiesel, or public procurement of fleet vehicles, such as flex-fuel vehicles for police forces or the post office to encourage the uptake of renewable fuels. PP affects a substantial share of world trade flows. For example, in the European Union this accounts for roughly 18% of GDP.

Pyrolysis. The breaking apart of complex molecules by heating in the absence of oxygen, producing solid, liquid and gaseous fuels.

Recycling. The processing of used waste materials into new products to reduce waste production, reduce the consumption of fresh raw materials, reduce energy usage, reduce air and water pollution and lower greenhouse gas emissions as compared to virgin production. Many materials, such as thermoplastics and glass, are recyclable. Although typically an alternative to landfilling, there is some debate about whether recycling is economically efficient. The costs and energy used in collection and transport compared to the costs and energy saved in the production process have been debated.

Renewable. In sustainable development, this relates to a commodity or resource, such as solar energy or firewood, that is inexhaustible or replaceable by new growth. The opposite is finite resources, such as fossil fuels, which are definitely exhaustible. The meaning is extended by including emissions. By dint of being replaceable within a relatively short time frame, renewable resources are more likely to be carbon-neutral as the emissions from their use can be negated by CO_2 capture during growth. As fossil resources take millions of years to develop, the emissions generated are considered "permanent" as the time frames nullify the original carbon capture (millions of years ago) in relation to the human lifespan.

Renewable Energy Directive (RED). Officially titled as 2009/28/EC, this is an EU directive that mandates levels of renewable energy use within the European Union. The directive was published on 23 April 2009 and amends and repeals the 2001 Directive on Electricity Production from Renewable Energy Sources. It requires the European Union to fulfil at least 20% of its total energy needs with renewables by 2020 – to be achieved through the attainment of individual national targets. All EU countries must also ensure that at least 10% of their transport fuels come from renewable sources by 2020. A new Renewable Energy Directive for the period after 2020 is in preparation.

Renewable energy tax credit. This is any tax credit offered by a local or federal taxation authority as an incentive for the installation and operation of renewable energy systems such as solar or wind power. Renewable power generation creates power in the form of electricity, and environmental benefits to society from "green" power production – such as minimising pollution and slowing the rate at which finite fuel resources are used. The electricity is sold into the local grid, and the societal benefits are sold in the form of renewable energy credits (RECs), sold separately as a commodity into the marketplace. For each REC purchased, customers can claim the equivalent MWh of energy reduction as an offset to their conventional energy use.

Renewable feedstock. This can be defined as any renewable, biological material that can be used directly as a fuel, or converted to another form of fuel, energy or bio-based material product. Biomass feedstocks are the plant and algal materials used to derive fuels like ethanol, butanol, biodiesel and other hydrocarbon fuels. Organic wastes are assuming greater importance politically as renewable feedstocks.

Renewable Fuels Standard (RFS). The RFS is a US federal programme that requires transportation fuel sold in the United States to contain a minimum volume of renewable fuels. The RFS originated with the Energy Policy Act of 2005 and was expanded and extended by the Energy Independence and Security Act of 2007 (EISA).

Salination. Soil salinity is the salt content in the soil; the process of increasing the salt content is known as salination or salinisation. Salts occur naturally within soils and water. Salination can be caused by natural processes such as mineral weathering or by the gradual withdrawal of an ocean. Salinity from irrigation can occur over time wherever irrigation occurs, since almost all water contains some dissolved salts. Soil salinity has detrimental effects on plant growth and yield.

Secondary refining. Further conversion and processing steps create a larger number of products from the intermediates. Thereby, in a first conversion step, the intermediate materials are fully or partially processed into precursors, as well as into more intermediates;

in further value creation at the site of the biorefinery, these are then fully or partially refined into products. The products from biorefineries can be both finished or semi-finished.

Slash. The component within the forest residues generated from sawlog processing typically consisting of chunks, foliage, branches and other broken material not appropriate to be comminuted by a chipper.

Small to medium-sized enterprise (SME). In the European Union, an SME can be defined as a firm with revenues of EUR 10-50 million or a balance-sheet total of EUR 10-43 million. In general, an SME has up to about 250 employees. SMEs are very common in biotechnology, especially in research-intensive firms. Many firms are involved in industrial biotechnology and attempt to make sustainable alternatives to fossil-derived goods.

Soil destruction. Soil destruction can include soil erosion. Soil can also be destroyed by salination, over-fertilisation and industrial pollution. If, for example, a soil becomes so contaminated with heavy metals from industry that it cannot be used in agriculture, it would be considered "destroyed", although the soil itself remains.

Soil erosion. Soil erosion is defined as the wearing away of topsoil. Topsoil is the top layer of soil and is the most fertile because it contains the most organic, nutrient-rich materials. Therefore, this is the layer that farmers want to protect for growing their crops and animals. Soil erosion can have several causes. A prime concern in sustainability is erosion caused by deforestation.

Solvent. This is a liquid, typically other than water, used for dissolving other substances. Often solvents are non-polar liquids that are toxic to humans and pose threats to the environment if released accidentally. Another major function of solvents is to clean surfaces, and biodegradable solvents have been developed as potential replacements for more harmful solvents in such applications.

Specialty chemicals. Single molecules or mixtures valued for their particular abilities – for example, killing bacteria or fire retardation.

Starch. A molecule composed of long chains of glucose molecules linked together (repeating unit $C_{12}H_{16}O_5$). This polysaccharide is widely distributed in the vegetable kingdom and is stored in all grains and tubers.

Starch biorefinery. The component separation in primary refining results in starch, which thus constitutes the platform of the starch biorefinery. Typical feedstocks are cereals or potatoes.

Steam explosion. This is a pre-treatment process in which biomass is treated with hot steam under pressure followed by an explosive decompression of the biomass that results in a rupture of the biomass fibres rigid structure, literally "exploding" the biomass to pulp. It makes the biomass polymers more accessible for subsequent processes, such as fermentation, hydrolysis or densification.

Stover. The dried stalks and leaves of a crop that remain after the grain has been harvested.

Sugar biorefinery. Sucrose, commercially available sugar, results from separation processes, and the sugar is then converted, usually through fermentation, to products such as ethanol. Typical feedstocks are sugar cane and sugar beet.

Supply chain. A network of individuals or organisations that procures materials; transforms these materials into intermediate and finished products; and distributes finished products to customers.

Sustainable development goals (SDGs). These are an inter-governmental set of 17 aspiration goals with 169 targets. The goals are contained in paragraph 54 of the United Nations Resolution A/RES/70/1, of 25 September 2015. They are officially known as Transforming our World: the 2030 Agenda for Sustainable Development. They include ending poverty and hunger, improving health and education, making cities more sustainable, combating climate change, and protecting oceans and forests.

Sustainability. This term is in common use, but is hard to define. Human and ecological sustainability have become intermingled with the emergence of climate change as a major societal challenge. Defining sustainability as a part of the concept of sustainable development, the Brundtland Commission of the United Nations in 1987 stated that: "sustainable development is development that meets the needs of the present without compromising the ability of future generations to meet their own needs". It reflects the realisation that economic growth must be achieved with minimal environmental damage if the effects of climate change are to be controlled.

Sustainability indicators. Ecological and environmental indicators have been intermingled in assessing sustainability. One subset of environmental indicators is ecological indicators that include physical, biological and chemical measures such as atmospheric temperature or the concentration of ozone in the stratosphere. These are also referred to as "state" indicators as their focus is on the state of the environment or conditions in the environment. A second subset is indicators that measure human activities or anthropogenic pressures, such as greenhouse gas emissions. These are also referred to as "pressure" indicators, i.e. they measure the pressures that humans place on the environment. Finally, there are indicators, such as the number of people served by sewage treatment, which track societal responses to environmental issues.

Synthesis gas (syngas). The product of gasification i.e. CO, H_2 and CO_2, a mixture of which is in itself a fuel.

Synthesis gas biorefining. Here there is no separate component separation during primary refining; instead, all organic constituents (e.g. solid domestic waste) and biomass components are broken down in such a way to produce the raw product synthesis gas. This makes materials amenable for biorefining that would otherwise be unsuited. Products range from fuels, such as Fischer-Tropsch diesel and methanol, to higher alcohols and chemicals, and chemicals.

Synthetic biology. The application of science, technology and engineering to facilitate and accelerate the design, manufacture and/or modification of genetic materials in living organisms. It aims to bring an engineering approach to biotechnology by design and engineering of biologically-based parts, novel devices and systems, as well as redesigning existing, natural biological systems. It has clear and current applications to bio-production of chemicals, plastics, textiles and other materials.

Thermochemical conversion. The use of heat to chemically change substances to produce energy products.

Torrefaction. Torrefaction of biomass is a mild form of pyrolysis at temperatures typically between 200°C and 320°C. It changes biomass properties to provide a much better fuel quality for combustion and gasification applications. It leads to a dry product that is stable on storage as rotting can no longer occur. It can also result in much higher energy density, useful to improve transportation efficiency.

Transcriptomics. The transcriptome is the complete set of transcripts in a cell, and their quantity, for a specific developmental stage or physiological condition. Transcriptomics therefore is the study of genes being expressed at any given time under given conditions.

Transesterification. Biodiesel can be produced from straight vegetable oil (SVO), animal oil/fats, tallow and waste oils. SVO creates fairly severe engine problems such as poor atomisation. Transesterification is the reaction of a triglyceride (fat/oil) with an alcohol to form esters (the biodiesel) and glycerol. The glycerol is relatively simple to separate, and the biodiesel has much enhanced properties as a diesel fuel.

Value added. The additional value, in customer terms, created at a particular stage of production.

Value chain. The value chain identifies the various value-adding activities of an organisation or network. It is often used as a tool for strategic planning because of its emphasis on maximising value while minimising costs.

Vegetable oil biorefinery. A biorefinery in which the feedstock is oil from various seeds and fruits, whereby oil is present together with other lipids.

Waste. This means any substance or object which the holder discards or intends to, or is required to, discard. This narrow definition avoids the implication that the material or object serves no further use. In the context of sustainability, waste minimisation is an important concept. Billions of tonnes of organic waste materials are produced worldwide every year, and a great deal of these waste materials could be used as a source of biomass for the production of bio-based materials.

Waste hierarchy. Waste disposal legislation has introduced a hierarchy of options for managing wastes. It gives top priority to preventing waste in the first place. When waste is created, it gives priority to preparing it for reuse, then recycling, then other recovery such as energy recovery, and last of all disposal, e.g. landfill.

Waste-to-energy. The practice of processing waste products to generate steam, heat or electricity.

Wastewater and wastewater treatment. Wastewater is any water that has been adversely affected in quality by human influence. Wastewater can originate from a combination of domestic, industrial, commercial or agricultural activities, surface run-off or storm water, and from sewer inflow or infiltration. Wastewater treatment is the process whereby wastewater is treated to render it innocuous and/or reusable. Biological wastewater treatment plants, known more widely as sewage treatment plants, are more sustainable than non-biological processes and are deployed worldwide. Wastewater treatment and reuse is considered essential to future water security.

Water security. The capacity of a population to safeguard sustainable access to adequate quantities of acceptable quality water for sustaining livelihoods, human well-being, and socio-economic development, for ensuring protection against water-borne pollution and water-related disasters, and for preserving ecosystems.

Zero waste. The goal of developing products and services, managing their use and deployment, and creating recycling systems and markets to eliminate the volume and toxicity of waste and materials, and conserve and recover all resources. Implementing zero waste eliminates all discharges to land, water or air that may be a threat to planetary, human, animal or plant health.

ORGANISATION FOR ECONOMIC CO-OPERATION AND DEVELOPMENT

The OECD is a unique forum where governments work together to address the economic, social and environmental challenges of globalisation. The OECD is also at the forefront of efforts to understand and to help governments respond to new developments and concerns, such as corporate governance, the information economy and the challenges of an ageing population. The Organisation provides a setting where governments can compare policy experiences, seek answers to common problems, identify good practice and work to co-ordinate domestic and international policies.

The OECD member countries are: Australia, Austria, Belgium, Canada, Chile, the Czech Republic, Denmark, Estonia, Finland, France, Germany, Greece, Hungary, Iceland, Ireland, Israel, Italy, Japan, Korea, Latvia, Luxembourg, Mexico, the Netherlands, New Zealand, Norway, Poland, Portugal, the Slovak Republic, Slovenia, Spain, Sweden, Switzerland, Turkey, the United Kingdom and the United States. The European Union takes part in the work of the OECD.

OECD Publishing disseminates widely the results of the Organisation's statistics gathering and research on economic, social and environmental issues, as well as the conventions, guidelines and standards agreed by its members.

OECD PUBLISHING, 2, rue André-Pascal, 75775 PARIS CEDEX 16
(92 2018 01 1 P) ISBN 978-92-64-29233-8 – 2018

www.ingramcontent.com/pod-product-compliance
Ingram Content Group UK Ltd.
Pitfield, Milton Keynes, MK11 3LW, UK
UKHW051501290926
15471UKWH00008B/1343

9 789264 292338